Framework and Tools for Environmental Management in Africa

Authors

Godwell Nhamo is a Programme Manager for the Exxaro Resources Ltd sponsored Chair in Business and Climate Change hosted by the Institute for Corporate Citizenship (ICC) at the University of South Africa (Unisa). Dr Nhamo holds a PhD from Rhodes University and did his postdoctoral work with the University of Witwatersrand. Dr Nhamo has great interests in business and climate change as well as environmental management and policy. Some of Dr Nhamo's current responsibilities are in teaching, research and training in corporate citizenship, sustainability sciences as well as business and climate change.

Ekpe Inyang holds an MSc degree in Environmental Studies obtained from the University of Strathclyde, UK. Mr Inyang's early childhood experiences with the Korup rainforest communities seemed to have influenced his educational ambition and orientation and spurred him to become a powerful advocate of conservation. Mr Inyang has many years of experience, having worked in different capacities with the World Wide Fund for Nature (WWF-UK), Wildlife Conservation Society (WCS-US) and the Pan African Institute for Development West Africa (PAID-WA) in Cameroon. Mr Inyang has special interests in environmental and development issues, gender equality, good governance and research. He is author of eight full-length plays, a collection of poems, a number of scientific articles and a few textbooks. His other textbooks include *The Forest: An African Traditional Definition; Environmental Education in Theory and Practice; Doing Academic Research; Redefining African Development* (co-authored by Nana Célestin) and *Environmental Problems in the Bakossi Landscape*.

Framework and Tools for Environmental Management in Africa

Godwell Nhamo & Ekpe Inyang

CODESRIA

Council for the Development of Social Science Research in Africa

DAKAR

ISBN: 978-2-86978-321-8

Layout: Hadijatou Sy
Cover Design: Ibrahima Fofana
Printed by : Imprimerie Graphi plus, Dakar, Senegal
Distributed in Africa by CODESRIA
Distributed elsewhere by the African Books Collective, Oxford, UK.
Website: www.africanbookscollective.com

The Council for the Development of Social Science Research in Africa (CODESRIA) is an independent organisation whose principal objectives are to facilitate research, promote research-based publishing and create multiple forums geared towards the exchange of views and information among African researchers. All these are aimed at reducing the fragmentation of research in the continent through the creation of thematic research networks that cut across linguistic and regional boundaries.

CODESRIA publishes a quarterly journal, *Africa Development*, the longest standing Africa-based social science journal; *Afrika Zamani*, a journal of history; the *African Sociological Review*; the *African Journal of International Affairs*; *Africa Review of Books* and the *Journal of Higher Education in Africa*. The Council also co-publishes the *Africa Media Review*; *Identity, Culture and Politics: An Afro-Asian Dialogue*; *The African Anthropologist* and the *Afro-Arab Selections for Social Sciences*. The results of its research and other activities are also disseminated through its Working Paper Series, Green Book Series, Monograph Series, Book Series, Policy Briefs and the *CODESRIA Bulletin*. Select CODESRIA publications are also accessible online at www.codesria.org.

CODESRIA would like to express its gratitude to the Swedish International Development Cooperation Agency (SIDA/SAREC), the International Development Research Centre (IDRC), the Ford Foundation, the MacArthur Foundation, the Carnegie Corporation, the Norwegian Agency for Development Cooperation (NORAD), the Danish Agency for International Development (DANIDA), the French Ministry of Cooperation, the United Nations Development Programme (UNDP), the Netherlands Ministry of Foreign Affairs, the Rockefeller Foundation, FINIDA, the Canadian International Development Agency (CIDA), the Open Society Initiative for West Africa (OSIWA), TrustAfrica, UN/UNICEF, the African Capacity Building Foundation (ACBF) and the Government of Senegal for supporting its research, training and publication programmes.

Contents

Chapter 4

Chapter 5

Chapter 6

PART II: SELECTED TOOLS FOR ENVIRONMENTAL MANAGEMENT

Chapter 7

Chapter 8

Chapter 9
Understanding Environmental Education 45

Chapter 10
Sustainability Reporting 163

Chapter 11
Promotion of Formal and Non-formal Environmental Education 187

PART III: EMERGING ENVIRONMENTAL ISSUES
Chapter 12
Conclusion: Emerging Issues and the Way Forward 211

Acronyms

ACCA	Association of Chartered Certified Accountants
AICC	African Institute of Corporate Citizenship
AKASA	Awareness, Knowledge, Attitude, Skills, Action
AR	Afforestation and Reforestation
BCSD	Business Council for Sustainable Development
BEE	Black Economic Empowerment
BOD	Biological Oxygen Demand
CBOs	Community Based Organisations
CDM	Clean Development Mechanism
CFC	Chlorofluorocarbons
CH_4	Methane
CIDA	Canadian International Development Agency
CITES	Convention in International Trade in Endangered Species
CO	Carbon Monoxide
CoP	Conference of the Parties
CSR	Corporate Social Responsibility
DANIDA	Danish International Development Agency
DDT	Dichlorodiphenyltrichloroethane
DEAT	Department of Environment and Tourism
DESD	Decade of Education for Sustainable Development
DNA	Designated National Authority
DOE	Designated Operational Entity
EA	Environmental Appraisal
EA	Environmental Assessment – or – Environmental Auditing
EB	Executive Board
ECA	Economic Commission for Africa
ECR	Excellence in Corporate Reporting
EE	Environmental Education
EEASA	Environmental Education Association of Southern Africa
EIA	Ecological Impact Assessment – or – Environmental Impact Analysis – or – Environmental Impact Assessment
EMP	Environmental Management Plan

ENDA   Environment and Development Activities
EPI   Environmental Performance Index
ESI   Environmental Sustainability Index
FAO   Food and Agricultural Organisation
FTSE   Financial Times Stock Exchange
GDP   Gross Domestic Product
GEF   Global Environmental Facility
GIS   Geographic Information Systems
GRI   Global Reporting Initiative
HFCs   Hydrofluorocarbons
HIA   Health Impact Assessment
HIV-AIDS   Human Immunodeficiency Virus - Acquired Immunodeficiency Syndrome
IDSA   Institute of Directors of South Africa
IET   International Emissions Trading
IPCC   Intergovernmental Panel on Climate Change
ISR   Integrated Sustainability Reporting
IUCN   World Conservation Union
IWMI   International Water Management Institute
JI   Joint Implementation
JSE   Johannesburg Stock Exchange
MDGs   Millennium Development Goals
MEAs   Multilateral Environmental Agreements
MoMET   Ministry of Environment and Tourism
N_2O Nitrogen Oxide
NEAPs   National Environmental Action Plans
NEPAD   New Economic Partnership for Africa's Development
NGO   Non-governmental Organisation
NRM   Natural Resource Management
NTFP   Non-timber Forest Products
OAU   Organisation for African Unity
ODSs   Ozone-depleting Substances
Pb   Lead
PDD   Project Design Document
PFCs   Perfluorocarbons
RISDP   Regional Indicative Strategic Development Plan
RSA   Republic of South Africa
SADC   Southern African Development Community
SAICM   Strategic Approach to International Chemicals Management
SAIEA   Southern African Institute for Environmental Assessment

SF_6	Sulphur Hexafluoride
SIA	Social Impact Assessment
SIDA	Swedish International Development Cooperation Agency
SoER	State of Environment Reporting
SRG	Sustainability Eeporting Guidelines
SRI	Socially Responsible Investment
TBL	Triple Bottom Line
ToR	Terms of Reference
UN	United Nations
UNCCD	United Nations Convention to Combat Desertification
UNCED	United Nations Conference on Environment and Development
UNDP	United Nations Development Programme
UNEP	United Nations Environment Programme
UNESCO	United Nations Educational, Scientific and Cultural Organisation
UNFCCC	United Nations Framework Convention on Climate Change
US	United States of America
WCED	World Commission on Environment and Development
WMO	World Meteorological Organisation
WRI	World Research Institute
WSSD	World Summit on Sustainable Development
WWF	World Wide Fund for Nature

Preface

For a long time, Africa's environmental problems have been addressed out of context. This context can be split into two: the colonial and postcolonial periods. Although both of these contexts are worthy of equal attention, it is in the postcolonial era that well calculated strategies to prolong enslavement have been used deliberately by our former masters, leading to the continent's impoverishment. In addition, many studies on the environment have concentrated on science: evidence of environmental decay, resulting in an information gap in the frameworks and tools for good environmental stewardship.

This book, therefore, does not dwell much on repeating and re-presenting the science and the uncontested facts of environmental decay in the continent. Rather, we assume and maintain that enough facts on environmental degradation have already been documented. What is lacking is the need to understand how relevant policy frameworks and tools, available both at global and continental levels, can be utilised to reverse the current and possible prolonged decay in our environment. Discussions of frameworks and tools for environmental management in Africa cannot evade the role of international multilateral environmental agreements, nor other global landmarks, such as the Earth Summit and the Millennium Development Goals.

First, the conceptual framework for environmental management in Africa is presented. Selected historical landmarks and donor and aid agency narratives are outlined. The book then sets out selected key environmental tools, such as environmental impact assessment, public participation, environmental education, corporate governance and sustainability.

This textbook has been written against the background of short supply of effective and up-to-date educational material on environmental management available to African scholars, industry, educators, government officials and NGOs. Not only are these stakeholders disempowered by the shortage of environmental management material, they are critically short of African-oriented environmental management textbooks.

Godwell Nhamo
& Ekpe Inyang

Chapter 1

Introduction and Overview

African governments have laid their hands on most international proposals made in response to the need to address environmental decay. But, unlike the sequential and harmonious progression witnessed from the developed countries, starting from the Stockholm Declaration on Human Environment in 1972, through the Rio de Janeiro Earth Summit on Environment and Development in 1992, to the Johannesburg World Summit on Sustainable Development in 2002, Africa has been forced into a discrete progression in addressing the requirements of such global landmarks. Evidently, a number of factors have forced the continent into such positions, and they can be listed, among which are the need to be part of the global village, the quest for rapid economic development and attached donor strings that firmly established, and are prolonging, the master-slave relationship between Africa and its former colonial masters. Clearly, we have for a long time been beggars of resources from our former colonial and other masters. Through such set-ups, some African governments have been held to ransom and arm-twisted for the love of development aid that has done little to eradicate extreme poverty and environment damage.

The Concept of Environment

The issues pertaining to the framework and tools for good environmental stewardship in Africa may not be fully understood without analysing the concept of environment. In this book, the term environment is conceptualised as being constituted by both the following dimensions: biophysical (natural) and human (socio-economic and political) dimensions. The biophysical dimension is made up of elements such as climate (temperature, rainfall, wind and evaporation), air, topography, geology, soils, vegetation (flora), fauna (animals), groundwater (hydrogeology), and surface water (hydrology). On the other hand, the human dimension constitutes element such as people, land tenure and use, archaeological, social, cultural, political and economic aspects. However, both the biophysical and human environments should be viewed as constantly interacting in a dynamic nature that supports all forms of life on earth.

This kind of conceptualisation of the environment helps us understand issues surrounding the uncertainty of environmental stewardship in Africa. We are in a position to engage holistically with some of the discourses that ground and shape narratives pertaining to environmental threats, particularly those of non-African origin, and the motives behind them.

African Environments and Colonial Histories

It may be wrong to assume that the current environmental problems stem from African origin, neglect or lack of civilisation. Much of the current environmental decay is a manifestation of Africa's colonial histories. Such histories testify to the fact that the partitioning of the continent and its resources, including the brutal dissection of cultural arrangements and physical boundaries, have contributed immensely to the wounded terrain we see and experience today.

Populations were re-grouped into squashed, infertile and fragile ecological zones across the face of the continent, resulting in the collapse of ecosystems and the emergence of the cycle of poverty – environment degradation. With the aid of forced, near-slave labour, mineral wealth was exploited and exported with much greed. Scores of huge open pits and scary underground mine shafts are still visible today. Africans had to make ends meet, and sacrifice their lives to regain dignity and land through many liberation wars. Fertile soils and healthy ecological zones were reserved for the privileged few colonial masters who, even today, have vast tracts of land (including under-utilised land) which they refuse to release for equitable redistribution. Hence, the fact remains that Africa's environmental problems mirror its past. It is only when all stakeholders acknowledge these anomalies that a sustainable solution to the current continued environmental damage can be holistically understood and found. Yet, many of our former colonial masters wish to conceal this critical pillar of understanding that shapes our future. Instead, they portray a picture of failing African governments, ineffective governance structures, corruption, poverty, the HIV/AIDS scourge, and so on. Therefore, in this text, the environment is as much understood a political and social question as it is an environmental question. However, many donor agencies present different narratives.

A Struggle for Environmental Management Space

The manner in which various Northern donors and countries have placed the environment on African governments' agenda is best seen as a struggle for environmental management space in Africa; and similarly, as a contest and battle for political, social and economic realisation by the foreign entities involved. The possible results (giving the benefit of the doubt) are the unintended consequences of the re-partitioning of Africa through the so-called environmental strategies and the associated finance mechanisms originating outside the continent. Some of the key financiers and aid agencies are the World Bank, the Canadian International Development Agency (CIDA), Swedish International Development Agency (SIDA) and the Danish International Development Agency (DANIDA). These, along with other aid agencies and donors, have come up with a number of environment related strategies that include: CIDA's Policy for Environmental Sustainability in 1992; Strategy for Denmark's Environmental Assistance to Developing Countries 2004-8, SIDA's 2002 Country

Strategies: Guidelines for Strategic Environmental and Sustainability Analysis; and the World Bank's Environment Strategy. More recently, the Africa Union, African Development Bank and other sub-regional groupings have developed 'home grown' strategies for good environmental stewardship. The key provisions of some of these environmental strategies are revisited in Chapter Four. A number of international NGOs, such as Greenpeace, have also propelled various agenda for environmental management in Africa.

The Challenge

This book outlines a framework for environmental management in Africa. It synthesises tools available to address environmental decay on the continent. It aims to bring to our attention the discourses surrounding who shapes and decides the continent's environmental policies. It answers questions such as: In what contexts are such policies and tools designed and implemented? How are these policies and tools applied and addressed to solve the crucial environmental concerns and problems we face? Why are these environmental policies and tools failing, or why are they not producing the desired results, in and for Africa?

The text reveals that part of the problem has been the fact that in recent years, the development field in Africa has witnessed a proliferation of global discourses and a flurry of ideas and activities promising to tackle perennial problems, among them inequality, environmental degradation and underdevelopment. Whilst the global development community and the African continent are grappling with the conceptualisation and implementation of one discourse, another discourse is suddenly unveiled and thrown into the mix. This means that the development community, particularly in Africa, has to literally and immediately pursue the new idea, almost always restructuring existing research and policy programmes to fit new concepts. It then becomes an aspect of the 'old' versus the 'new' environment agenda, rather than governments giving themselves time to harmonise new ideas with old ideas, and finding a way forward. Many environmental policies and tools have suffered this dilemma, and environmental concerns and problems have been only partially addressed.

Until recently, modernisation theories emanating from the developed world have dominated development thought and environmental policies, leading to research and development interventions that ignore the indigenous and local knowledge bases and skills that Africa can offer. The marginalisation and the neglect of indigenous environmental knowledge, due to its perceived inferiority, has led to growing popularity and dominance of Western science, characterised by its so-called 'universalism'. Where appropriate, this text will provide insights with regard to the manner in which African governments can use the tools available to address environmental concerns and problems 'the African way'.

Book Outline

This book comes in two major parts: framework for environmental management (Part I) and selected tools for environmental management (Part II). The sub-themes addressed in the various chapters are summarised in the following paragraphs.

Chapter Two focuses on global landmarks for environmental management, including The Brundtland Report of 1987, the United Nations Summit on Environment and Development that took place in Rio de Janeiro, Brazil, 14–17 June 1992, leading to the adoption of Agenda 21 as the global action plan on sustainable development, the World Summit on Sustainable Development, the Millennium Development Goals and the Decade of Education for Sustainable Development.

Sources and fundamentals of international environmental law are discussed in Chapter Three, which presents law-making treaties, and how the procedures around negotiation, adoption, authentication and ratification are conducted. International customs and the general principles of environmental laws are considered. However, the larger part of the chapter is dedicated to the fundamentals of selected multilateral environmental agreements (MEAs), which are of significance to Africa, and not discussed elsewhere in the book. Some of the MEAs highlighted include the Convention on Biological Diversity, the Ramsar Convention, the Convention on International Trade of Endangered Species of Flora and Fauna, Convention to Combat Desertification, United Nations Framework Convention on Climate Change, and the Kyoto Protocol. The last part of the chapter dwells on addressing implementation gaps faced when domesticating these MEAs.

Chapter Four interrogates how environmental problems are put on the African agenda. The role of key donors such as SIDA, DANIDA and CIDA, and how their various environmental management strategies have pursued specific enviro-political agendas, are outlined. The chapter also traces how new Afro-centric enviro-political agendas are emerging with the coming into effect of the Africa Union's New Economic Partnership for Africa's development (NEPAD) Environment Strategy, and the African Development Bank Group's Policy on the Environment. The above chapters make up Part I.

Chapter Five is dedicated to addressing key environmental problems in the continent with an emphasis on common resources. Among the issues addressed are deforestation, wildlife de-population and extinction, land degradation, drought, desertification, soil impoverishment, air and water pollution, global warming, and gender and the environment.

Aspects relating to Africa and global environmental problems are addressed in Chapter Six. Some of the major global environmental problems discussed in much depth include air pollution, global warming and climate change, ozone-layer depletion and acid deposition. The chapter also addresses issues of nuclear waste and waste treatment in general.

Chapter Seven considers natural resources, and definitions and objectives of conservation and natural resource management. From the definition, it is established that conservation and natural resource management are synonymous. The chapter highlights some of the causes and effects of the mismanagement of natural resources. It analyses the distinction and link between conservation and sustainable development, advancing arguments and suggesting strategies for conservation. It ends by emphasising the need to seek the active participation of women in conservation and natural resource management, given their various levels of interaction with environmental resources.

Issues concerning environmental impact assessment (EIA) and public participation are considered in Chapter Eight. The chapter addresses these aspects in a unique fashion, looking at both the typical project and EIA cycles simultaneously. The value of this approach is the benefit that stakeholders using this resource will have when addressing sustainability issues at various project stages, especially with regard to the involvement of the public in practical set-ups.

Environmental education, as one of the key environmental management tools, is reviewed in Chapter Nine. The chapter starts with a historical account of the development of environmental education, highlighting the various movements and key actors. It proceeds to a discussion of the conferences that led to the formulation of its definition and objectives, ratification and internationalisation. Adequate attention is given to the definition, scope, forms and objectives of the subject. A clear distinction is made between awareness raising and sensitisation, in order to inform policy and practice. A discussion of organisational structures and institutions that provide excellent environments and programmes is aimed at schools, local communities and the general public. Some prerequisite activities, prior to the implementation of an environmental education programme, are also proposed.

Chapter Ten introduces the relatively new concept of sustainability reporting. This tool has been effective in making businesses conscious of their environmental, social and community responsibilities. Through various voluntary and quasi-mandatory initiatives, sustainability reporting has resulted in companies doing more for the environment in Africa than before. Instead of reporting and accounting only on financial matters, corporate entities are now able to report on ways in which the 'triple bottom line' of the economic, environmental and social aspects of sustainable development is being effectively addressed both within and outside company premises.

Chapter Eleven deals with presentations around non-formal and formal environmental education programmes, approaches for their implementation and methods, as well as monitoring and evaluation tools. These issues are usually given little space, or are neglected in some of the literature dealing with tools of good environmental management, especially in Africa.

The single-chapter contained in Part III, Chapter Twelve presents emerging environmental concerns and tools, and concluding remarks. Emphasis is on the need to consider climate change and the loss of biodiversity as the key environmental management challenges. To minimise the impacts of climate change and loss of biodiversity, including human security, African governments and individual communities are encouraged to take strong positions to work together to address the problems in an 'African way'.

This chapter re-introduces the key elements covered in this book. The term 'environment' captures both human and non-human facets. The term 'environment' embraces a holistic picture. It is not used in a limited context to cover conservation or ecology alone. The chapter explains the fact that it is wrong to attribute African environmental ills to political, social and economic neglect by African governments. But rather, it needs to be acknowledged that colonial histories still remain embedded in the terrain of the continent. This kind of acknowledgement is necessary to give those in positions of influence, power and authority a holistic overview, putting in place strategies to address environmental decay in the continent.

PART I

FRAMEWORK FOR ENVIRONMENTAL MANAGEMENT

Chapter 2

Global Landmarks for Environmental Governance

Introduction

The 'environment' became a global policy issue in the mid-1960s (Carter 2001). By that time, many governments had adopted a techno-centric approach that considered environmental problems to be the unfortunate side-effects of economic growth and development. Therefore, the main assumption was that governments would eventually find a way of addressing such problems (Howlett and Ramesh 1995). The standard approach to dealing with such environmental problems was re-active rather than pro-active (UNEP 2002; UNEP 2003b; UNEP 2003a). This approach could not, however, stem the ever-increasing and complex environmental problems, such as resource depletion, waste, pollution and global warming.

A number of global landmarks in environmental management and policy are worth mentioning: The World Commission on Environment and Development (WCED), which produced the now famous document *Our Common Future* (or The Brundtland Report) (WCED 1987), the United Nations Summit on Environment and Development that took place in Rio de Janeiro, Brazil, 14–17 June 1992, leading to the adoption of Agenda 21 as the global action plan on sustainable development (UNCED 1992), the World Summit on Sustainable Development (UN 2002), the Millennium Development Goals (UNDP 2003) and the Decade of Education for sustainable Development (UNESCO 2004). These and other relevant landmarks are discussed in the following sections.

The Stockholm Declaration of 1972

The Stockholm Declaration was made in Stockholm, Sweden in June 1972. This followed an invitation by Sweden, as it had just experienced 'severe damage to thousands' of its lakes from acid rain following critical air pollution in Western Europe (UNEP 2003a:4). Twenty-six principles and an Action Plan of 109 recommendations guide the Stockholm Declaration (UNEP 2003a; Sands 2003; Leeson 1995). The general principles from the Stockholm Declaration are presented in Box 2.1.

Box 2.1: Principles of the Stockholm Declaration

Principle 1:	Human rights must be asserted, apartheid and colonialism condemned
Principle 2:	Natural resources must be safeguarded
Principle 3:	The earth's capacity to produce renewable resources must be maintained
Principle 4:	Wildlife must be safeguarded
Principle 5:	Non-renewable resources must be shared and not exhausted
Principle 6:	Pollution must not exceed the environment's capacity to clean itself
Principle 7:	Damaging oceanic pollution must be prevented
Principle 8:	Development is needed to improve the environment
Principle 9:	Developing countries therefore need assistance
Principle 10:	Developing countries need reasonable prices for exports to carry out environmental management
Principle 11:	Environment policy must not hamper development
Principle 12:	Developing countries need money to develop environmental safeguards
Principle 13:	Integrated development planning is needed
Principle 14:	Rational planning should resolve conflicts between environment and development
Principle 15:	Human settlements must be planned to eliminate environmental problems
Principle 16:	Governments should plan their own appropriate population policies
Principle 17:	National institutions must plan development of states' natural resources
Principle 18:	Science and technology must be used to improve the environment
Principle 19:	Environmental education is essential
Principle 20:	Environmental research must be promoted, particularly in developing countries
Principle 21:	States may exploit their resources as they wish but must not endanger others
Principle 22:	Compensation is due to states thus endangered
Principle 23:	Each nation must establish its own standards
Principle 24:	There must be cooperation on international issues
Principle 25:	International organisations should help to improve the environment
Principle 26:	Weapons of mass destruction must be eliminated

Source: Adopted from Clarke and Timberlake (1982) as cited by UNEP (2003: 3).

The significant achievements of the Stockholm resolutions were: (a) recommendations for the establishment of new institutions and coordinating mechanisms for the institutions already in place (the Action Plan); (b) the definitions of a framework for future actions to be undertaken by the international community (the recommendations); and (c) the adoption of the guiding principles outlined in Box 2.1 (Sands

2003). The Action Plan recommended that the UN General Assembly formulate four institutions that included an inter-governmental Governing Council for Environmental Programmes to guide and coordinate environmental management programmes; an environment secretariat; an Environment Fund and lastly, an inter-agency Environmental Co-ordinating Board that would ensure cooperation and coordination among major bodies involved in the implementation of environmental programmes within the UN systems. The United Nations Environment Programme (UNEP) was born out of the Stockholm Declaration as the secretariat. Today, its mission is spelt out as the need to:

> Provide leadership and encourage partnership in caring for the environment by inspiring, informing and enabling nations and peoples to improve their quality of life without compromising that of future generations (UNEP 2003a:4).

According to Clarke and Timberlake (cited in UNEP 2003a:5), the environment was placed firmly on most government agendas following the Stockholm Declaration. The authors indicate that there were only around ten ministries and departments responsible for the environment in existence prior to the declaration. Whereas by 1982, more than 110 countries had established such ministries or departments to deal with pressing environmental matters.

Sandbrook (1992), credits the Stockholm Conference for having managed to place the environment in the global arena. However, he notes deep divisions between countries of the North (developed) and those of the South (emerging and developing), an aspect that re-surfaced again during the Rio Summit. According to Sandbrook's observations, the conference was a dialogue of the deaf between the rich and the poor. In order to clean up the polluted world, governments from the North advocated for all nations and industry to agree to share the burden. However, governments of the South wanted industry to create more jobs and eradicate poverty, even at the cost of the environment. In the perspective of former India's Prime Minister Indhira Ghandi, of all the pollutants that were faced, the worst was poverty (Sandbrook 1992).

Our Common Future and Sustainable Development

Our Common Future called upon world governments to embrace the concept of sustainable development, defined as development that 'meets the needs of the present without compromising the ability of future generations to meet their own needs' (WCED 1987:8). This way, sustainable development implied capturing the three conventional pillars namely: economic, social and environmental (Figure 2.1).

The aim of the World Commission on Environment and Development was to find practical ways of addressing the environmental and developmental problems of the world (WCED 1987). In particular, it had three general objectives: re-examine the critical environmental and development issues and formulate realistic proposals for dealing with them; propose new forms of international cooperation on these issues so as to influence policies and events in the direction of needed

Figure 2.1: Sustainable Development from *Our Common Future*

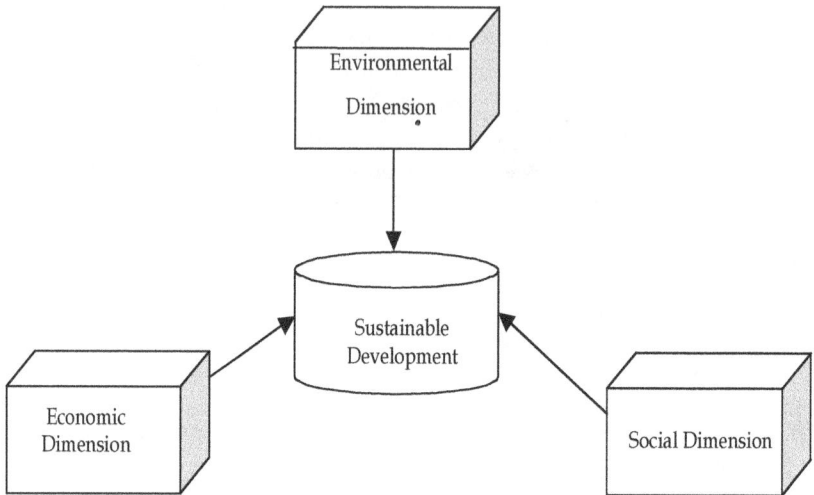

changes; and raise the levels of understanding and commitment to action of individuals, voluntary organisations, businesses, institutes and governments. *Our Common Future* reported on many global realities and recommended urgent action on eight key issues to ensure that development was sustainable. These were: population and human resources, food security, the urban challenge, energy, industry, species and ecosystems, managing the commons, and conflict and environmental degradation. These eight key issues were identified as early indicators of sustainable development.

Unsustainable development was attributed mostly to the limitations of technology and social organisation, natural resources, and the ability of the biosphere to take up the cumulative negative impacts from human activities. Hence, both technology and social organisation could be regulated and improved to pave the way for a new era of sustainable economic growth (WCED 1987). Unsustainable economic growth was understood as the main cause of skewed resource distribution (including income) leading to poverty, as people failed to provide basic needs such as food, shelter and clothing. The poor were associated with the direct impoverishment of the environment, as 'a world in which poverty is endemic will always be prone to ecological and other catastrophes' (WCED 1987:8). In this regard, calls were made for economic growth that realised the importance of the earth's life support systems: water, soil and the atmosphere (Cahill 2002).

Our Common Future's definition of sustainable development was also adopted during the UN Earth Summit of 1992, which set Agenda 21 as a global action plan for implementing sustainable development (UNCED 1992). However, in as much as the definition takes cognisance of people, many environmental policies and legislation, both on global and national scales, it hardly recognises them as the primary focus of development (Jacobs cited in Cahill 2002:2). In this respect, Jacobs identifies equity (commitment to meet basic needs of the poor), quality of life (that economic growth should not be taken to equate to human well-being), and participation (involving as many stakeholders as possible in environmental policy processes) as additional key themes in attaining sustainability. Cahill (2002) also warns of the need to distinguish between the concepts of sustainability and sustainable development. He maintains that the former refers to the end-state, whereas the latter refers to the means by which that end is achieved. However, the two terms are often used interchangeably. This is the case in this book.

Although popular and, at times, generic, the phrase 'sustainable development' is complex and often highly contested. This has resulted in its mis-use at various fora, from the grassroots, through to national, regional and global levels. In fact, it has become a global refrain. Therefore, what we present here are a few pointers to what we believe could comprise true sustainability in managing the environment for poverty eradication in Africa. True sustainability implies that we as individuals, households, communities, nations and the whole world at large succeed in removing selfish motives from development; are ready to be good environmental stewards; and that world governments set the right platforms for true dialogue and let sustainability agenda filter down to regional and national levels, resulting in tangible deliverables at the grassroots. The last point means refraining from repeating extraordinarily resource-intensive global talk-shows, such as the 1992 Rio Summit, 1997 Rio+5 Summit and the 2002 World Summit on Sustainable Development held in Johannesburg, South Africa, commonly known as the WSSD, and its Implementation Plan.

Leeson (1995) advocates the principle of sustainable development. He is particularly concerned with broader national or regional trends, and the long-term consequences of negative social and economic developments. This confronts policymakers with the task to select between 'more immediate, quantifiable merits of a proposed course of action and the more speculative benefits to future generations of present self-denial' (Lesson 1995:38). However, in Africa, the time has come when we may need to consider sustainable development and sustainability as open questions. Since no one can really accurately predict the future, we should continue searching for appropriate positioning and responses, and approach sustainability with an open mind, acting responsibly.

The Rio Declaration on Environment and Development

The Rio Declaration on Environment and Development, Rio de Janeiro, 3–14 June 1992, underscored twenty-seven principles (UN 1992). These principles, shown in Box 2.2, are set out here in full; bearing in mind that access to online information remains critical in many African countries. The principles will permit further debate and research into how far the African continent and those outside it have measured up to them.

Box 2.2: Principles Established by the Conference on Environment
and Development, 1992

Principle 1: Human beings are at the centre of concerns for sustainable development. They are entitled to a healthy and productive life in harmony with nature.

Principle 2: States have, in accordance with the Charter of the United Nations and the principles of international law, the sovereign right to exploit their own resources, pursuant to their own environmental and developmental policies, and the responsibility to ensure that activities within their jurisdiction or control do not cause damage to the environment of other States or of areas beyond the limits of national jurisdiction.

Principle 3: The right to development must be fulfilled so as to equitably meet developmental and environmental needs of present and future generations.

Principle 4: In order to achieve sustainable development, environmental protection shall constitute an integral part of the development process and cannot be considered in isolation from it.

Principle 5: All States and all people shall cooperate in the essential task of eradicating poverty as an indispensable requirement for sustainable development, in order to decrease the disparities in standards of living and better meet the needs of the majority of the people of the world.

Principle 6: The special situation and needs of developing countries, particularly the least developed and those most environmentally vulnerable, shall be given special priority. International actions in the field of environment and development should also address the interests and needs of all countries.

Principle 7: States shall cooperate in a spirit of global partnership to conserve, protect and restore the health and integrity of the earth's ecosystem. In view of the different contributions to global environmental degradation, States have common but differentiated responsibilities. The developed countries acknowledge the responsibility that they bear in the international pursuit of sustainable development, in view of the pressures their societies place on the global environment and of the technologies and financial resources they command.

Principle 8: To achieve sustainable development and a higher quality of life for all people, States should reduce and eliminate unsustainable patterns of production and consumption and promote appropriate demographic policies.

Principle 9: States should cooperate to strengthen endogenous capacity-building for sustainable development by improving scientific understanding through the exchange of scientific and technological knowledge, and by enhancing the development, adaptation, diffusion and transfer of technologies, including new and innovative technologies.

Principle 10: Environmental issues are best handled with the participation of all concerned citizens, at the relevant level. At the national level, each individual shall have appropriate access to information concerning the environment that is held by public authorities, including information on hazardous materials and activities in their communities, and the opportunity to participate in decision-making processes. States shall facilitate and encourage public awareness and participation by making information widely available. Effective access to judicial and administrative proceedings, including redress and remedy, shall be provided.

Principle 11: States shall enact effective environmental legislation. Environmental standards, management objectives and priorities should reflect the environmental and developmental context to which they apply. Standards applied by some countries may be inappropriate and of unwarranted economic and social cost to other countries, in particular developing countries.

Principle 12: States should cooperate to promote a supportive and open international economic system that would lead to economic growth and sustainable development in all countries, to better address the problems of environmental degradation. Trade policy measures for environmental purposes should not constitute a means of arbitrary or unjustifiable discrimination or a disguised restriction on international trade. Unilateral actions to deal with environmental challenges outside the jurisdiction of the importing country should be avoided. Environmental measures addressing transboundary or global environmental problems should, as far as possible, be based on an international consensus.

Principle 13: States shall develop national law regarding liability and compensation for the victims of pollution and other environmental damage. States shall also cooperate in an expeditious and more determined manner to develop further international law regarding liability and compensation for adverse effects of environmental damage caused by activities within their jurisdiction or control to areas beyond their jurisdiction.

Principle 14: States should effectively cooperate to discourage or prevent the relocation and transfer to other States of any activities and substances that cause severe environmental degradation or are found to be harmful to human health.

Principle 15: In order to protect the environment, the precautionary approach shall be widely applied by States according to their capabilities. Where there are threats of serious or irreversible damage, lack of full scientific certainty shall not be used as a reason for postponing cost-effective measures to prevent environmental degradation.

Principle 16: National authorities should endeavour to promote the internalisation of environmental costs and the use of economic instruments, taking into account the approach that the polluter should, in principle, bear the cost of pollution, with due regard to public interest and without distorting international trade and investment.

Principle 17: Environmental impact assessment, as a national instrument, shall be undertaken for proposed activities that are likely to have a significant adverse impact on the environment and are subject to a decision of a competent national authority.

Principle 18: States shall immediately notify other States of any natural disasters or other emergencies that are likely to produce sudden harmful effects on the environment of those States. Every effort shall be made by the international community to help States so afflicted.

Principle 19: States shall provide prior and timely notification and relevant information to potentially affected States on activities that may have a significant adverse transboundary environmental effect and shall consult with those States at an early stage and in good faith.

Principle 20: Women have a vital role in environmental management and development. Their full participation is therefore essential to achieve sustainable development.

Principle 21: The creativity, ideals and courage of the youth of the world should be mobilized to forge a global partnership in order to achieve sustainable development and ensure a better future for all.

Principle 22: Indigenous people and their communities and other local communities have a vital role in environmental management and development because of their knowledge and traditional practices. States should recognise and duly support their identity, culture and interests and enable their effective participation in the achievement of sustainable development.

Principle 23: The environment and natural resources of people under oppression, domination and occupation shall be protected.

Principle 24: Warfare is inherently destructive of sustainable development. States shall therefore respect international law providing protection for the environment in times of armed conflict and cooperate in its further development, as necessary.

Principle 25: Peace, development and environmental protection are interdependent and indivisible.

Principle 26: States shall resolve all their environmental disputes peacefully and by appropriate means in accordance with the Charter of the United Nations.

Principle 27: States and people shall cooperate in good faith and in a spirit of partnership in the fulfilment of the principles embodied in this Declaration and in the further development of international law in the field of sustainable development.

Source: Report of the United Nations Conference on Environment and Development, Rio de Janeiro, 3–4 June 1992.

The Rio summit stipulated one of its fundamentals as the 'precautionary principle' (Principle 15). The precautionary principle states that 'action to protect the environment against the danger of severe and irreversible damage need not wait for rigorous scientific proof' (Weiss 2003:137). It transpired because most issues of environmental policy depend on science and technology. The precautionary principle has led to governments developing strategies for pro-active management of environmental risks under conditions of scientific uncertainty, which is usually the situation in decisions affecting environmental policy (Weiss 2003). Other principles of interest, with regard to recent environmental management trends in Africa, are the application of economic instruments, especially the 'polluter pays' principle (Principle 16), and the adoption and application of environmental impact assessments

(EIAs) to ensure development does not harm the environment (Principle 17). As presented in Chapter Seven, many countries in Africa now have legislation to guide EIAs.

During Rio, the divide between developed and developing economies persisted. Governments from the North had an agenda to 'solve issues of the climate, forests and endangered species' (Sandbrook 1992:16). For governments from the South, it was the same story of poverty. They lobbied for the coupling of development and the environment. Hence, if the North wanted the South to stop deforestation, slow down the consumption of fossil fuels (chiefly coal) and reduce birth rates, then the North had to pay.

The Rio conference witnessed the adoption of Agenda 21, a blueprint for sustainable development (UNCED 1992). Agenda 21 (21st Century Agenda) is a list of action points agreed by world governments. The plan realised that economic development neglects other developmental issues (Castro 2004). Agenda 21 actions include: promoting environmentally sound management of solid waste and sewage; combating poverty; education, training and awareness; protecting and promoting human health; protecting the atmosphere; managing fragile ecosystems; and conserving biological diversity. Since then, many multilateral environmental agreements (see Chapter Three) have been put in place and ratified by many countries globally, including South Africa. A full review of the progress in addressing Agenda 21 was undertaken during the World Summit on Sustainable Development (WSSD) held in Johannesburg, South Africa, in September 2002, which led to the WSSD Plan of Implementation (UN 2002).

Article 21 of the WSSD Plan of Implementation calls for the prevention and minimisation of waste, and the maximisation of re-use, recycling, and the use of environmentally friendly alternative materials. It also calls for the participation of all government authorities and stakeholders in minimising adverse effects on the environment. Actions to achieve this include the implementation of a waste management hierarchy.

The desire to promote education, public awareness and training (Chapter 36 of Agenda 21) in order to achieve good environmental stewardship was recognised as one of the key means of implementing Agenda 21. Chapter 36 makes reference to the Declaration and Recommendation of the Tbilisi Intergovernmental Conference on Environmental Education, which was held in 1977 and provided the fundamental principles. Three key issues and the basis for action were spelt out (Quarrie 1992). They include:

Reorienting Education Towards Sustainable Development

Education was to be recognised as a process by which societies can reach their fullest potential, thereby improving capacity to address environmental and development related issues. To be effective and efficient, environmental and developmental education was supposed to deal with the dynamics of both the natural and human environments. These were supposed to be integrated into all disciplines, making use of both formal and non-formal methods, as well as effective means of communication.

Increasing Public Awareness

Chapter 36 noted the low levels of awareness about the interrelated nature of anthropogenic activities and the environment. This was linked to inaccurate and insufficient information, particularly for developing countries. In addition, developing countries also faced problems related to technology and environmental expertise.

Promoting Training

This should have a job-specific focus, aimed at filling identified gaps regarding the knowledge base and necessary skills in environmental and development management. Simultaneously, training must promote awareness of environmental and development concerns (as a two-way process) at local, national and global levels.

UN Conference on Desertification

The United Nations Conference on Desertification (UNCOD), held in Nairobi in 1997, remains a global landmark in addressing environmental concerns of desertification. An estimated 500 delegates from ninety-four countries gathered in Nairobi in August and September 1997 to discuss the problems of desertification. UNCOD spelt out the Plan of Action, the immediate objective of which was to prevent and arrest the advance of desertification and, where possible, to reclaim desertified land for productive use (UN 1997). The main objective was to sustain and promote, within ecological limits, the productivity of arid, semi-arid, sub-humid and other areas vulnerable to desertification in order to improve the quality of life of inhabitants. The Plan of Action outlined twenty fundamental principles. UNCOD acknowledged that countries affected by desertification are at different stages with respect to their appreciation of and ability to address desertification problems. Given this scenario, countries were required to first define the extent and impact of desertification by: strengthening or establishing a national body for assessment and monitoring of desertification; and determining criteria for identifying and assessing desertification and its causes. If the problem of desertification existed, a system to monitor the problem would be set up (UN 1997).

Millennium Development Goals

Millennium Development Goal 7 stipulates the need to ensure that environmental sustainability is achieved at the lowest possible scale, thus, the household (UNDP 2003). Three targets were set, namely to: integrate the principle of sustainable development into national policies and programmes by 2015; halve the proportion of people without access to safe drinking water and basic sanitation; and by 2020, achieve significant improvement in the standards of living of at least 100,000,000 squatter residents. The risk of urban squatters remains. Fragile and even stable ecosystems easily succumb to heavy population densities, resulting in the depletion of naturally occurring life support systems, and which affect quality of life issues like waste management and sanitation. Other goals are stipulated as to: eradicate extreme poverty and hunger; achieve universal primary education; promote gender equality and empower women; reduce child mortality; improve maternal health; and combat HIV/AIDS, malaria and other diseases.

The MDG environmental objectives, unlike the others, have been criticised for their generic and immeasurable targets (International Atomic Energy Agency 2005). In addition, the MDGs also fail to take cognisance of possible synergies across the goals. One area African governments should work towards, if implementation is to be accelerated, is harmonising and domesticating the MDGs. However, there are other challenges associated with the domestication of MDGs. Priority areas differ. For example, Zimbabwe has placed as a priority the need to work towards social development and poverty eradication (Government of Zimbabwe 2004). Zimbabwe prioritises the MDGs addressing poverty, the empowerment of women and HIV/AIDS. It therefore came as no surprise that the country became the first African nation to have a female vice president in 2005. Although not a priority area, the Ministry of Environment and Tourism was tasked with achieving the MDG environmental goals (Box 2.3).

Box 2.3: Zimbabwe's Environmental Targets and other Issues for MDGs

Targets

- integrate the principles of sustainable development into national policies and programmes and reverse the loss of environmental resources
- halve, by 2015, the proportion of people without access to safe drinking water and basic sanitation
- by 2020, achieve a significant improvement in the housing condition of at least one million slum dwellers, peri-urban and high-density lodgers.

Indicators

- proportion of land area covered by forest
- land area protected to maintain biological diversity
- GDP per unit of energy use (as proxy of energy efficiency)
- proportion of people with sustainable access to improved water source
- proportion of people with access to improved sanitation
- number of housing units produced annually.

Challenges

- implementation of the land resettlement programme in a sustainable manner
- provision of decent housing in urban areas
- provision of safe water and sanitation, particularly in rural areas

- establish waste management practices to combat air and water pollution
- implementation of the provisions in the 2002 Environmental Management Act
- implementation of multilateral environmental agreements
- energy provision.

Priorities for Development
- environmental awareness
- strengthen development of appropriate alternative renewable energy sources
- provision of decent housing in urban sanitation programme
- consolidation of the rural water supply and sanitation programme
- improved management of urban environments
- expand biodiversity.

Priority for Development Assistance
- implementation of multilateral environmental agreements
- environmental awareness
- capacity building in data collection and analysis.

Source: Compiled from Government of Zimbabwe (2004:51–4).

These are generic priorities for most African countries. For example, the South African municipal election campaigns prior to 1 March 2006 were filled with messages around service delivery and housing. The bucket system became a burning issue, the ANC-led government having failed to eliminate it in a decade. The land question was also deliberated, with a number of political groupings questioning the potential of the willing-buyer-willing-seller arrangement to solve critical land distribution imbalances that favour the former settler masters. Other issues of concern emerging during the campaign included water supply and energy.

Another point of interest (Box 2.3) concerns priorities for development assistance. For a long time, the African continent has been a 'cry baby'. It is time we forge ahead with continental partnerships and share expertise, for example in capacity building for reliable data collection and analysis. Leaders should prioritise funding for research. We are fed with statistics of various kinds, with degrees of bias, due to the fact that Africa does not have coherent databases on various statistics (HIV/AIDS included). Hence, it is important to note that in many cases, statistics are as reliable as the purposes to which they are being manipulated. To this end, we may have to accept that some of our environmental management decisions could have been based on wrong statistics and were therefore wrong.

Decade of Education for Sustainable Development

UNESCO (2004) realises that the concept of sustainable development has changed and will continue to do so. As such, in pursuing education for sustainable development, the Decade of Education for Sustainable Development (DESD), which began in 2005, presents three dimensions of sustainable development: society, environment and economy. Culture has an underlying dimension. Society, the environment and the economy are thus defined, respectively, as:

> An understanding of social institutions and their role in change and development, as well as the democratic and participatory systems which give opportunity for the expression of opinion, the selection of governments, the forging of consensus and the resolution of differences.

> An awareness of the resources and fragility of the physical environment and the effects on it of human activity and decisions, with a commitment to factoring environmental concerns into social and economic policy development.

> A sensitivity to the limits and potential of economic growth and their impact on society and on the environment, with a commitment to assess personal and societal levels of consumption out of concern for the environment and for social justice (UNESCO 2004:4).

To this end, DESD's global vision is spelt out as 'a world where everyone has the opportunity to benefit from quality education and learn the values, behaviour and lifestyles required for a sustainable future and for positive societal transformation' (UNESCO 2004:23). The key actors are absorbed into the DESD the moment they accept it, thereby becoming stakeholders in the process. Three sets of stakeholders and their roles were identified: governmental and inter-governmental bodies, civil society and NGOs, and the private sector. Governmental and inter-governmental bodies are responsible for policy-formulation, promoting public consultation and input, conducting national and international public campaigns, and integrating education for sustainable development into education systems. Civil society is responsible for public awareness raising, advocacy and lobbying, consultancy and input into policy formulation, executing DESD in non-formal set-ups, participatory learning and action, as well as for mediation between governments and people. The private sector was given responsibility for entrepreneurial initiatives and training, management models and approaches, implementation and evaluation, and the development and sharing of practices of sustainable production and construction. Five key objectives for DESD were framed (UNESCO 2004:4):

- give an enhanced profile to the central role of education and learning in the common pursuit of sustainable development

- facilitate links and networking, exchange and interaction among stakeholders in education for sustainable development

- provide a space and opportunity for refining and promoting the vision of, and transition to sustainable development – through all forms of learning and public awareness

- foster increased quality of teaching and learning in education for sustainable development

- develop strategies at every level to strengthen capacity in education for sustainable development.

DESD is characterised as holistic, and as being interdisciplinary in its approaches to learning for sustainable development across the curriculum. It is values-driven, promotes critical thinking and problem solving, employs multi-methods, is participatory in decision making, and promotes the need to address locally relevant issues as part of the global platform.

Seven strategies are outlined for DESD: advocacy and vision building, consultation and ownership, partnership and networks, capacity building and training, research and innovation, information and communication technologies, and monitoring and evaluation. The outcomes from DESD are measured by the changed lives of 'thousands of communities and millions of individuals as new attitudes and values inspire decisions and actions, making sustainable development a more attainable ideal' (UNESCO 2004:5).

State of Environment Reporting

State of environment reporting (SoER) has become one of the key policy tools for spearheading good environmental governance in Africa. SoER is supposed to provide regular environmental updates by various levels of government. National SoER initiatives usually take place at five-year intervals (MoMET 1998). Five years is considered an adequate period over which 'significant' environmental change can be observed. However, this period is also recommended, as it can be easily adjusted to coincide with election and government cycles in many African states.

Over the years, SoER has become more structured, following an integrated approach to environmental assessment, with a reporting framework having emerged (UNEP 2003a). The reporting framework seeks to establish the causal relationship between humans and nature. It outlines the relationships between causes (now commonly cited as drivers and pressures) to environmental outcomes (the state), and to activities (policies and decisions) that shape the environment and its transformations. SoER is intended to cover the following major themes and major sub-themes: social-economic trends, land, forests, biological diversity, freshwater, coastal and marine, atmosphere, urban areas and disasters. Details of the themes and their sub-themes for SoER within the African context are shown in Box 2.4.

Box 2.4: Key Themes in State of Environment Reporting

Land:Consider the levels of degradation and desertification as well as inappropriate and inequitable land tenure

Forests:Record levels regarding deforestation and loss of forest quality

Biodiversity: Look at habitat degradation and loss and the bushmeat trade

Freshwater: Describe and measure vulnerability of water resources, water stress and scarcity, access to safe water and sanitation, deteriorating water quality as well as wetlands loss

Coastal and Marine: Document coastal area erosion and degradation, levels of pollution as well as climate change and sea-level rise

Atmosphere: Assess air quality, climate variability and vulnerability to climate change as well as floods and drought

Urban Areas: Search for signs of rapid urbanisation, assess waste quantities, determine levels of water supply and sanitation as well as air pollution

Disasters: Record events such as droughts, floods, armed conflict and earthquakes.

Source: Compiled from UNEP 2003:31

Depending on the level at which SoER takes place, details and the depth of reporting increase from the continental to the sub-continental, to the national and provincial, and ultimately to the local. Although state of the environment reports have been developed, particularly at sub-regional and national levels in Africa, challenges in the manner in which information is formulated and the regularity with which the reports are produced remain. Most reports are highly technical and scientific. This makes them inaccessible to many policymakers: many of our parliamentarians have only attained basic education status. The reports have had limited impact on policy development and implementation. These aspects require serious redress. A few countries, including South Africa, have realised this limitation and have started producing policy-oriented briefs around various issues pertaining to environmental management in the country packaged in 'policy-friendly' language (DEAT 2002).

Conclusions

This chapter has discussed fundamentals concerning global landmarks. Some of the landmarks elaborated include: the Stockholm Declaration, The World Commission on Environment and Development that produced the now famous document *Our Common Future* (The Brundtland Report), the United Nations Summit on Environment and Development that took place in Rio de Janeiro, Brazil from 14-17 June 1992 leading to the adoption of Agenda 21 as the global action plan on sustainable development, the World Summit on Sustainable Development, the Millennium Development Goals and the Decade of Education for Sustainable Development. The chapter has also conceptualised sustainable development.

Revision Questions

1. What are the fundamental principles of the Stockholm and Rio Declarations?
2. Which major thematic areas should be captured when reporting on the state of the environment?
3. What are the key provisions of the Millennium Development Goals?

Critical Thinking Questions

1. How practical and applicable is *Our Common Future's* conceptualisation of sustainable development to the African environmental agenda?
2. What measures has your government put in place to address the United Nations Decade for Sustainable Development?
3. From your assessment, are these measures adequate?
4. If not, what alternatives could be suggested to improve on the situation?
5. What measures should be put in place to quicken the pace with which the African Union could achieve the Millennium Development Goals?

References

Cahill, M., 2002, *The Environment and Social Policy*, London: Routledge.

Carter, N., 2001, *The Politics of the Environment: Ideas, Activism and Policy*, Cambridge: Cambridge University Press.

Castro, C.J., 2004, 'Sustainable Development: Mainstream and Critical Perspectives', in *Organisation and Environment*, Vol. 17, pp. 195–225.

DEAT, 2002, *Guideline on Recycling Solid Waste*, Pretoria: Government Printer.

Government of Zimbabwe, 2004, *Zimbabwe Millennium Development Goals: 2004 Progress Report*, Harare: Government Printer.

Howlett, M. and Ramesh, M., 1995, *Studying Public Policy: Policy Cycles and Policy Subsystems*, New York: Oxford University Press.

International Atomic Energy Agency, 2005, *Energy Indicators for Sustainable Development: Guidelines and Methodologies*, Vienna: International Atomic Energy Agency.

Leeson, J.D., 1995, *Environmental Law*, London: Pitman Publishing.

MoMET, 1998, *Zimbabwe's State of the Environment '98*, Harare: Government Printers.

Quarrie, J., 1992, 'Agenda 21', in J. Quarrie, ed., 1992, *Earth Summit '92: The United Nations Conference on Environment and Development Rio de Janeiro 1992*, London: The Regency Press.

Sandbrook, R., 1992, 'From Stockholm to Rio', in J.Quarrie (ed.), *Earth Summit '92: The United Nations Conference on Environment and Development Rio de Janeiro 1992*, London: The Regency Press, pp. 15–17.

Sands, P., 2003, *Principles of International Environmental Law* (13th edition), Cambridge: Cambridge University Press.

UN, 1992, *Report of the United Nations Conference on Environment and Development*, New York: United Nations Secretariat.

UN, 2002, *World Summit on Sustainable Development Plan of Implementation*, New York: UN Secretariat.

UNCED, 1992, *Agenda 21*, New York: UN Secretariat.

UNDP, 2003, *Human Development Report 2003: Millennium Development Goals – A Compact among Nations to End Human Poverty*, New York: Oxford University Press.

UNEP, 2002, *Global Environment Outlook 3*, New York: Wiley.

UNEP, 2003a, *Global Environment Outlook 3: Past, Present and Future Perspectives*, London: Earthscan.

UNEP, 2003b, *Industry and the Environment, UNEP Environment Brief*, No. 7, Kenya: Nairobi, UNEP [pamphlet].

UNESCO, 2004, *United Nations Decade of Education for Sustainable Development 2005-2014*, Paris: UNESCO.

WCED, 1987, *Our Common Future*, Oxford: Oxford University Press.

Weiss, C., 2003, 'Scientific Uncertainty and Science-based Precaution', in *International Environmental Agreements: Politica, Law and Economics*, Vol. 3, pp. 137–66.

Chapter 3

International Environmental Law: Sources and Fundamentals

Introduction

The earth is plagued with a huge array of environmental problems, due largely to anthropogenic causes. These problems can be categorised as either local or global, depending on the scale or geographical spread of their impact. Though local problems are largely the concerns of nations, because of the apparent localised impact, it is now understood that sooner or later, such impacts will escalate and spread beyond national boundaries and may eventually assume proportions of global concern.

Awareness raising and sensitisation campaigns have often been employed to address environmental problems and concerns through changing attitudes. However, this strategy normally takes a long time to produce expected results. Often, extensive damage to the environment results. Therefore, the application of the law to address environmental problems, before they get out of proportion, is an imperative. Shaw (1997) cited Principle 24 of the Stockholm Declaration of 1972 as stating that international matters concerning the protection and improvement of the environment should be handled in a cooperative spirit. Principle 7 of the Rio declaration of 1992 emphasised the need for states to cooperate in a spirit of global partnership to conserve, protect and restore the health and integrity of the earth's ecosystems.

Cunningham et al. (2003) define laws as rules set by authority, society or custom. Environmental law is defined as a special body of official rules, decisions and actions concerning environmental quality, natural resources and ecological sustainability. Its purpose is to regulate human behaviour and activity in order to prevent worsening situations, in line with Umozurike's (1995) observation that the absence of rules is an invitation to chaos and anarchy.

Addressing major environmental problems requires international effort, mainly through international conventions and treaties. This chapter therefore examines the sources of international environmental law. It discusses some important international environmental conventions that have been put into force.

Sources of International Environmental Law

Glahn (1970), Hughes (1992), Umozurike (1995) and Shaw (1997) present elaborate discussions and analyses of the sources of international law, which are also applicable to international environmental law. They identify four sources of international law: law-making treaties, international customs, general principles of law and written texts. Each of these are discussed further in the following sections.

Law-making Treaties

Treaties, known by a variety of names, ranging from conventions, international agreements, pacts, general acts and charters, through to statutes, refer to written agreements whereby partaking states bind themselves legally to act in a particular way to establish particular relations between them (Shaw 1997). Treaties can simply be defined as agreements between two or more states that seek to establish relationships between themselves, governed by international law (Umozurike 1995).

Treaties may arise in a number of ways, for example because of pressure from a state, groups of states or an international organisation (Hughes 1992). There are various types of treaties; only law-making ones are sources of international environmental law.

Law-making treaties are concluded between a number of countries acting in their own interests, with the intention of creating new rules that are adhered to later by other states, either through formal actions in accordance with the provisions of the treaties, or by tacit acquiescence in observance of the new rules (Glahn 1970). Such types of treaties are instruments through which a number of states declare their understanding of particular rules of law, which establish new general rules governing the future conduct of ratifying or adhering states; abolish or modify some existing customary or conventional rules of law, or create new international agencies.

In view of the sovereign nature of modern states, such treaties are initially binding only on states that sign and ratify them. If the initial number of ratifying states is small, the treaties do not create new rules of general international law. At best, only rules of particular or regional application are created. However, as acquiescence to the new laws, or formal ratification of them by additional states increases, and finally, when an overwhelming majority of all states accepts the new rules, they become part of general international law.

Although treaties are considered to be an effective and reliable source of international environmental law (Umozurike 1995; Shaw 1997), it is important to note that they may take several years to come into force. They normally take about three stages, as described by (Glahn 1970):

Negotiation

Treaties are normally drawn up through a process of negotiation coordinated by any authorised person or organisation. Diplomatic or other official channels may be utilised, a meeting of representatives may be arranged, or an international conference may be convened for the purpose.

Adoption and Authentication

Once the text of a treaty has been drafted in a formal form during the negotiation, adoption by the parties takes one of several courses: mutual consent, in the case of bilateral agreement; unanimous consent, in the case of treaties negotiated between a limited number of states; by the voting rules adopted by the conference, in the case of multilateral instruments negotiated by an international conference, or according to the voting rules provided either by the constitution of the organisation or by the organ or agency competent to issue such rules, in the case of treaties drawn up in an international organisation, or at a conference convened by such an organisation.

Authentication of the treaty is achieved when negotiators initial the text on behalf of their states; the text is incorporated into the final act of the conference at which it was made; and the text is incorporated into a resolution adopted by an organ of an international organisation, or negotiators append their signatures to the text of the agreement.

Ratification

A majority of modern international treaties become effective only on ratification. Virtually every state has developed detailed domestic regulations outlining the process of treaty ratification, but there are certain commonalities. The process is generally held to be an executive act, undertaken by the head of state of the government, through which the formal acceptance of the treaty is proclaimed. Until such acceptance is proclaimed, a treaty does not create obligations for the state in question, except in some rare instances where an agreement becomes effective by signature alone. Most ratification processes involve discussions in national assemblies (parliaments), and in some instances by senates, depending on the provisions of a state's constitution. The parliament or senate then mandates the head of state, or any other designated authority, to ratify the instruments on behalf of the state.

International Customs

Customs represent a second source of international environmental law (Glahn 1970; Umozurike 1995). In contrast to the normal meaning of the term, notably the description of a habit, a legal custom represents usage with a definite obligation attached to it.

The presence of customary international law is evident from the existence of an extensive body of detailed rules, which comprised the bulk of accepted general international law until shortly after the end of the nineteenth century (Glahn 1970). Most of the rules governing such diverse rules as jurisdiction over territory, freedom of the high seas, privileges and immunities of states, and the rights of aliens fall into this sphere of law.

Some of the rules in question originated through the practices of a small number of states, which were adopted by other states because of their usefulness, until at last, general acceptance resulted in new rules of law entailing definite obligations. This is in line with Shaw's (1997) assertion that customs depend on a particular activity by one state being accepted by another state or states as an expression of a legal obligation or right. This is amply illustrated by Glahn (1970) who observes that a custom results from the existence of a single powerful nation in the West, which imposed its will on its neighbours in relation to certain matters. Eventually, other countries accepted the policy or practice, without challenge or protest. When the number of assenting states reached a near universal proportion, a new rule of law had been created.

Although customary international law is relevant in environmental issues, its impact is generally weak, because it recognises the principles of state or territorial sovereignty: that is the right of states to carry out activities for their own benefit. These include, for example, the right to develop industries and to carry out peaceful nuclear activity (Hughes 1992).

General Principles of Law

The statute ranks general principles of law recognised by civilised nations as the third source of international law (Umozurike 1995). General principles do not apply when there are relevant treaties or customs, but are simply called in to fill the gaps in the law so that the court is not incapacitated from giving a judgement, *non-liquet*. For example, a situation may arise where the court is considering a case whereupon it realises that there is no law, parliamentary statute or judicial precedent, covering a point exactly (Shaw 1997). In such instances the judge simply deduces a rule from already existing rules will be relevant, by analogy, or directly from the general principles that guide the legal system.

General principles therefore constitute a reservoir from which the courts may draw in appropriate cases, while recognising the dynamics of international law and the creative function of the courts in interpreting it (Glahn 1970). Early writers also drew inspiration from general principles, such as from Roman-Dutch law, in particular, the substantive, procedural and evidentiary aspects common to legal systems, and existing in both municipal and international laws. These principles are not applied as points of reference in passing judgments in the courts, but are useful guides in the application of the law.

Written Texts

Written texts, considered as subsidiary law determining agencies (Umozurike 1995), are another important source of international law. As Shaw (1997) states, the influence of academic writers on the development of international law is marked. In the heyday of natural law, the juristic opinions and critical analyses by academics were of crucial importance, while the role of state practice and court decisions were of less significance. However, the importance attached to the text depends on the prestige of the author and the extent to which the author's opinions withstand the test of time. Such writers, referred to by the statute as writers 'of various nations' (Glahn 1970), rise above national, racial and other subjective or prejudicial considerations. They are guided by objective reasoning and analyses. Based on these selection criteria, writers such as Gentilis, Grotius, Pufendorf, Bynkershoek, and Vattel stand out as the supreme authorities from the sixteenth to the eighteenth century. They determined the scope, form and content of international law. However, with the rise of positivism and the consequent emphasis on state sovereignty, treaties and customs assumed the dominant position in the exposition of the rules of the international system, to the extent that the importance attached to legalistic writing began to decline (Shaw 1997). This decline in importance notwithstanding, there are some textbook writers who have continued to exert tremendous impact on the evolution of some aspects of international law, for example Gidel on the law of the seas, and Oppendeim and Rousseau, whose general works on international law tend to be virtually referred to as classics.

Multilateral Environmental Agreements

As recognition of the interconnections in the global environment has advanced, the willingness of nations to enter into protective treaties has grown concomitantly (Cunningham et al. 2003). Often, governments sign these international agreements for prestige, to be part of the international community, and to avoid criticism. These agreements provide useful tools that could be employed by the international community to press for change. NGOs often use these same tools in their campaigns. Usually, violating signatory countries face the embarrassment of having to explain their actions at regular meetings. This could be incentive enough for them to consider changing. However, since treaties are only binding on signatory countries, and enforcement often relies on self-policing, only good faith can restrain nations prone to violation. This suggests that even signatory governments need some strong political will, and not just force, to evade the embarrassment of regularly giving explanations for violations, and to genuinely respect the terms of treaties.

International efforts to protect the natural environment can be dated to the 1870s. Switzerland first tried to establish a regional agreement to protect nesting sites of migratory birds (French 1992). In 1886, a convention was signed between Germany, The Netherlands, Luxembourg and Switzerland to regulate salmon fisheries. But it was not until the 1970s that the move to internationalise environmental policymaking gained serious momentum.

Governments have signed over 200 environmental treaties, covering subjects of shared concern, which include ocean pollution, endangered species, acid rain, habitat loss, hazardous waste production and export, climate change, biodiversity decline and sustainable development (UNEP 1996). A few of these conventions are discussed below.

Convention on Biodiversity

The earth's biological resources are vital to humanity's economic and social development. There is growing recognition that biological diversity is a global asset of tremendous value to present and future generations (UNEP 2003). Despite this wide recognition, there is an alarming increase in the rate of species extinction, due to the continued execution of environmentally unsafe human activities.

In response to the growing impact on the biodiversity of uncontrolled human activities, the United Nations Environmental Programme (UNEP) convened an Ad Hoc Working Group of Experts on Biological Diversity in November 1988 to explore the need for an international convention to address the problem. Shortly afterwards, in May 1989, it established an Ad Hoc Working Group of Technical and Legal Experts to prepare an international legal instrument for the conservation and sustainable use of the earth's biological resources. As part of the terms of reference, the groups considered sharing the costs and benefits, resulting from biodiversity conservation, between developed and developing nations, as well as ways and means of supporting innovations by local people.

The effort culminated in the signing of the biodiversity convention by 168 countries at the 1992 United Nations Conference on Environment and Development (UNCED), popularly known as the Earth Summit, in Rio de Janeiro, Brazil. Article 1 of the convention spells out that 'the objectives of this convention, to be pursued in accordance with its relevant provisions, are the conservation of biological diversity; the sustainable use of its components and the fair and equitable sharing of the benefits arising out of the utilisation of genetic resources, including by appropriate access to genetic resources and by appropriate transfer of relevant technologies, taking into account all rights over those resources and to technologies, and by appropriate funding'.

Article 3 of the convention highlights the provision of the Charter of the United Nations and principles of international law which stipulate that states have 'the sovereign right to exploit their own resources in pursuant of their environmental policies, and the responsibility to ensure that activities within their jurisdiction or control do not cause damage to the environment of other states or of areas beyond the limits of national jurisdiction'.

Article 10 provides that, as far as possible, each contracting party shall integrate considerations of the conservation and sustainable use of biological resources into national decision-making; adopt measures relating to the use of biological resources to avoid or minimise adverse impacts on biological diversity; protect and encourage customary use of biological resources in accordance with traditional cultural practices that are compatible with conservation or sustainable use requirements; support local populations to develop and implement remedial action in degraded areas where biological diversity has been reduced, and encourage cooperation between its governmental authorities and its private sector in developing methods for sustainable use of biological resources.

The convention focused principally on the mechanisms to finance sustainable biodiversity, its use and protection, and designated the existing Global Environmental Facility (GEF) as the institutional structure responsible for carrying out its provisions. GEF is a major Trust Fund, established in 1991, as a joint venture of the World Bank, the United Nations Development Programme (UNDP) and the United Nations Environmental Programme (UNEP). One of the goals of GEF is to provide funds to facilitate sustainable development and the conservation of natural resources. The convention also highlighted the importance of indigenous knowledge in conservation.

Convention on Wetlands of International Importance

Both fresh- and saltwater wetlands are critical areas of biological productivity. Therefore, wetlands are not only ecologically important, but are also economically significant sites. Unfortunately, these sites are faced with serious problems, including catchment, destruction, pollution, and reclamation. It became necessary for countries to commit to ensuring that their destruction is halted through means of an international convention.

The Convention on Wetlands is an intergovernmental treaty adopted on 2 February 1971 in the Iranian city of Ramsar, on the southern shore of the Caspian Sea. It has come to be known popularly as the 'Ramsar Convention'. The convention entered into force in 1975. As of August 2007, it had 155 contracting parties, or member states, in all parts of the world (Ramsar 2007a).

The first obligation under the convention is for a party to designate at least one wetland, at the time of accession, for inclusion in the List of Wetlands of International Importance (the 'Ramsar List'). In addition, the party should continue to 'designate suitable wetlands within its territory' for the list (Article 2.1), the selection of which is based on their significance in terms of ecology, botany, zoology, limnology or hydrology (Ramsar 2007a). The addition of a site to the Ramsar list confers upon a party the prestige of international recognition, and expresses the government's commitment to take all necessary steps to ensure the maintenance of the ecological character of the site.

However, the convention provides that a contracting party may, because of its 'urgent national interests', delete or restrict the boundaries of a wetland already included in the list (Article 2.5), but that such deletions or restrictions should be compensated for by the designation of another wetland with similar habitat values, either in the same area or elsewhere as a Ramsar site (Article 4.2). In practice, only a handful of boundary restrictions have occurred. The only sites ever deleted from the Ramsar list (with three new sites designated in compensation) were judged on the basis that they did not meet any of the criteria that were actually set after their designation, so instead, new sites were designated. As of 1 September 2007, the 155 contracting parties had designated 1,675 sites for the Ramsar list, covering an area of 150,200,000 million hectares (1,502,000 km^2) (Ramsar 2007b).

There is a general obligation for contracting parties to include wetland conservation considerations in their national land-use planning. This commitment is to ensure 'the wise use of wetlands in their territory' (Article 3.1). Parties are obliged to promote their conservation. It is important to note that Ramsar is the first of the modern global intergovernmental treaties on the conservation and sustainable use of natural resources. It is equally important to note that although UNESCO serves as Depositary for the Convention, the Ramsar convention is not part of the UN and UNESCO system of environment conventions and agreements. The convention operates in close cooperation with five NGOs: BirdLife International, the International Water Management Institute (IWMI), Wetlands International, the World Conservation Union (IUCN), and the World Wide Fund for Nature (WWF) (Ramsar 2007a).

Convention on International Trade in Endangered Species

The Convention on International Trade in Endangered Species of Flora and Fauna (CITES) was signed by twenty-one nations in 1973. It was intended to regulate the explosive growth in the trade of endangered plant and animal species, at both national and international levels. CITES has three appendices, which determine the restriction placed on trade in each of the endangered species. Article II of the conventions (UNEP 2003) spells out that Appendix I shall include all species threatened with extinction, which are or may be affected by trade. Trade in specimens of these species must be subject to particularly strict regulation in order not to further endanger their survival. It must only be authorised in exceptional circumstances. Appendix II includes all species, which although not necessarily now threatened with extinction may become so, if trade in specimens of such species is not subjected to strict regulation in order to avoid utilisation incompatible with their survival. Other species must be subject to regulation in order that trade in specimens of certain species may be brought under effective control. Appendix III includes all species which any party identifies as being subject to regulation within its jurisdiction for the purpose of preventing or restricting exploitation, and as needing the cooperation of the other parties in the control of trade.

Convention to Combat Desertification

Desertification is one of the most serious environmental problems in Africa, with implications for food security. Though desertification affects the African continent the most – two-thirds of the continent is desert or drylands – it is not a problem confined to drylands in Africa. It is a worldwide problem, directly affecting 250,000,000 people and a third of the earth's land surface, or over four billion hectares.

The issue of desertification was discussed globally at the UN Conference on Desertification held in Nairobi in 1977. Attempts to efficiently tackle the problem were crippled due to a lack of both administrative and financial support. Therefore in 1992, the United Nations Conference on Environment and Development

(UNCED) or the so-called Rio Earth Summit recommended the elaboration of a United Nations Convention to Combat Desertification (UNCCD) (www.unccd.entico.com/english/faq.htm). UNCCD was adopted in June 1994 and was opened for signature in October 1994 in Paris. It entered into force in December 1996, three months after the receipt of its fiftieth ratification. The convention is based on the principles of participation, partnership and decentralisation, the backbone of good governance, with the objective to:

> ...combat desertification and mitigate the effects of drought in countries experiencing serious drought and/or desertification, particularly in Africa, through effective action at all levels, supported by international cooperation and partnership arrangements, in the framework of an integrated approach which is consistent with Agenda 21, with a view to contributing to the achievement of sustainable development in affected areas (UNEP 1996).

The convention formulates a range of obligations on both affected country parties and on developed country parties. These include the following as general obligations: adopting an integrated approach addressing the physical, biological and socio-economic aspects of desertification and drought; giving due attention to the situation of affected developing country parties with regard to international trade, marketing arrangements and debt integrating strategies for poverty eradication into efforts to combat desertification; promoting cooperation among affected country parties in the fields of environmental protection and the conservation of land and water resources; strengthening sub-regional, regional and international cooperation; cooperation within relevant intergovernmental organizations; and determining institutional mechanisms and promoting the use of existing financial mechanisms and arrangements (UNEP 1996).

UNCCD now has more than 180 country parties, making it truly global in reach. UNCCD has reached maturity and is evolving from the preparation of National Action Programmes to their implementation. An assessment of programmes by the parties in 2000 and 2001 showed that the capacity strengthening efforts for key actors at the local level was successful in identifying and addressing challenges linked to sustainable development. Furthermore, the bottom-up approach of UNCCD helped to strengthen relationships between governments and local communities, particularly in larger countries. It also favoured the decentralised involvement of stakeholders and end users of natural resources in the development process (www.unccd.entico.com/english/faq.htm). UNCCD spells out that in order to achieve the objective, long-term integrated strategies would be employed 'that focus on improved productivity of land, and the rehabilitation, conservation and sustainable management of land and water resources'.

Convention on Substances that Deplete the Ozone Layer

The continuous use of substances that deplete the ozone layer, such as chlorofluorocarbons (CFCs), is seriously threatening life on earth. The ozone layer is the only atmospheric shield that protects the earth from the destructive rays of the sun. By 1985, the ozone layer had become so depleted that a hole had developed over Antarctica. This discovery was based on extensive research, the results of which were widely publicised in international media.

A series of meetings were held to discuss the issue. There was strong support for an international convention to be signed to address the problem. This convention, also known as the Montreal Protocol, was signed in 1987, adding specific obligations to the rather vague framework treaty known as the Vienna Convention. The protocol was strengthened in 1990. It set strict timetables for the phasing out of CFCs and other ozone-depleting substances (ODSs) by the year 2000. By July 1991, over seventy nations, thirty of which were from the developing world, had signed the convention. However, in recognition of their circumstances, developing nations were given a ten-year period of grace to reach full compliance with the terms of the protocol.

An interesting aspect of the convention, as stated in Article 10a, is that each party shall take every practicable step, consistent with the programmes supported by the financial mechanism, to ensure first that the best available, environmentally safe substitutes and related technologies are expeditiously transferred to parties operating under Paragraph 1 of Article 5; and second, that these transfers occur under fair and most favourable conditions (UNEP 2000).

Another important element is specifying controlled substances and products containing some of them in annexes. Annexes A, B, C, and E provide lists of controlled substances, indicating their ozone-depleting potentials. Annex D provides a list of products containing controlled substances specified in Annex A, including their custom code numbers. This provides the basis for proper identification, information, education and enforcement.

Convention on Climate Change

The signing of the convention on climate change was motivated by strong concerns that human activities had been substantially increasing the atmospheric concentrations of greenhouse gases; that these increases enhance the natural greenhouse effect; and that this will result on average in an additional warming of the earth's surface and atmosphere, with potential adverse effects on natural ecosystems and humankind (UN 2005). Over and above providing many goods and services that sustain rural livelihoods, promote environmental quality and advance economic development, natural resources serve as a first-line, defence against climate change (Agrawala et al. 2005). Projected carbon dioxide (CO_2) emissions up to 2030 are shown in Figure 3.1.

Figure 3.1: Projections of CO_2 Emissions

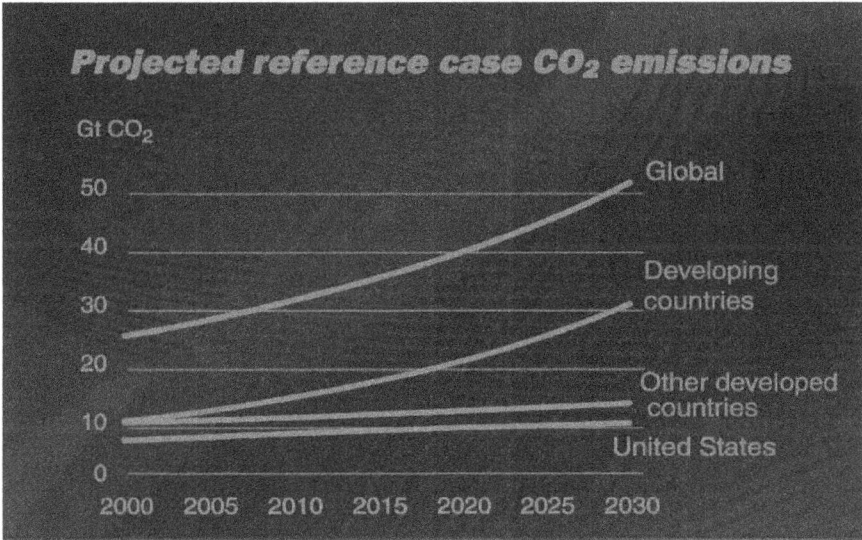

Source: Abareconomics (undated)

In spite of amounting evidence in support of climate change, this issue has remained one of the most controversial in international negotiations since the 1980s (Shimada 2004). While assessments of past and present emission patterns strongly influence debates over international climate policy, the central challenge is to limit future emissions (Baumert and Pershing 2004). Projections of future emissions are highly uncertain, particularly for developing countries.

United Nations Framework Convention on Climate Change

In 1988, the World Meteorological Organisation (WMO) and the United Nations Environmental Programme (UNEP) established the Intergovernmental Panel Climate Change (IPCC) to assess relevant information on climate change, covering impacts, adaptation and mitigation (Inriani 2005). A global agreement to mitigate climate change was proposed, culminating in the United Nations Framework Convention on Climate Change (UNFCCC). This convention was signed in 1992 at the massively attended United Nations Conference on Environment and Development in Rio de Janeiro, Brazil. Article 2 of the convention states: 'The ultimate objective of this convention and any related legal instruments that the Conference of the Parties may adopt is to achieve, in accordance with the relevant provisions of the convention, stabilisation of the greenhouse gas concentrations in the atmosphere at a level that would prevent dangerous anthropogenic interference with the climate system. Such a level should be achieved within a time frame sufficient to allow ecosystems to adapt naturally to climate change, to ensure that food production is not threatened and to enable economic development to proceed in a sustainable manner.'

The convention took into account some the specific needs and circumstances of developing countries, and the fact that these countries are particularly vulnerable to the adverse effects of climate change. These countries would have to bear a disproportionate and abnormal burden. To this effect, developed nations were asked to take the lead in combating climate change. But all parties were obliged to cooperate towards the achievement of the objective. However, the convention did not contain concrete plans to attain this objective (Shimada 2004).

The Kyoto Protocol

A Conference of the Parties (CoP) to the UNFCCC is held at least once a year. At the third CoP in 1997 in Kyoto, Japan, the Kyoto Protocol was adopted. This defined policies to reduce greenhouse gas emissions (Idriani 2005). Annex A of the protocol lists six greenhouse gases: carbon dioxide (CO_2), methane (CH_4), nitrous oxide (N_2O) sulphur hexafluoride (SF_6), hydrofluorocarbons (HFCs), and perfluorocarbons (PFCs) (UN 2005). The core strategy for reducing greenhouse gas emissions is to reduce the burning of fossil fuels by using them more efficiently, a policy that was met with strong opposition. Opponents of the protocol argue that that this would cause net economic damage, without acceptance of the immense benefits that Canada, for instance, has enjoyed from thirty years of energy conservation.

An essential part of the Kyoto Protocol is its 'flexibility mechanisms' (Idriani 2005). These comprise international emissions trading (IET), by which industrialised countries (referred to as Annex I countries) can trade part of their emission budgets between themselves; joint implementation (JI), which allows industrialised countries to earn emission credits from emission reduction projects in other Annex I countries; and the Clean Development Mechanism (CDM), which permits industrialised countries to gain emission credits from emission reduction projects in developing countries.

It is apparent that only the CDM gives developing countries the opportunity to be directly involved in the implementation of the protocol, with direct implications for meeting the emission reduction targets of Annex I countries. It is important to note that the establishment of the CDM under the Kyoto Protocol has been greeted with mixed emotions from climate, forestry and development experts (Manguiat et al. 2005). Many methodological and generic issues regarding the implementation of CDM activities, and the role of CDM in climate protection and national development, remained unsolved up CoP II, held in Montreal Canada in 2005, including the role of afforestation and reforestation (AR) projects as a specific part of CDM (Box 3). Given that natural resource-related projects are also major activities supported by official development assistance, the relationship between development assistance funding, CDM and AR remains somewhat opaque. The procedure in the implementation of CDM is quite complicated, as can be seen from a summary by Manguiat et al. (2005):

- project participants have to select an approved baseline and monitoring methodology or develop a project-specific one, which then needs to be approved by the CDM Executive Board (EB) (the international body in charge of supervising the CDM);

- in parallel, they produce a 'project design document' (PDD) that explains and assesses the planned activity according to a given scheme and applies the methodology in practice;

- the PDD and methodology are then submitted to the 'Designated National Authority' (DNA, which is the new domestic institution required, among other things, to implement the CDM in the host country);

- the DNA will issue a letter of endorsement to confirm that the project is contributing to the host country's sustainable development priorities;

- the correct application of the methodology and the consistency of the PDD will then be *validated* by the 'Designated Operational Entity' (DOE) (i.e. auditing companies and the like) and then *registered* by the EB;

- the registration is the prerequisite for a project to actually produce marketable emission reductions.

The Basel and Bamako Conventions

African countries, as is the case of many other low-income countries of the world, have remained easy targets for the dumping of hazardous wastes. This is a serious problem facing the continent and other parts of the developing world, which necessitated the signing of the Basel and Bamako Conventions.

The Basel Convention, signed in 1989 by both industrialised, and some 115 developing countries, has the objective to:

> ...set up obligations for State Parties (or signatory states) with a view to: (a) reducing transboundary movements of wastes subject to a minimum consistent with the environmentally sound and efficient management of such wastes, (b) minimising the amount and toxicity of hazardous wastes generated and ensuring their environmental sound management (including disposal and recovery operations) as close as possible to the source of generation; (c) assisting developing countries in environmentally sound management of hazardous and other wastes they generate (UNEP 1996:359).

UNEP (1996) provides a summary of the general provisions of the convention, which include: a) parties prohibiting the import of hazardous wastes shall inform other parties of their decision and for these other parties not to permit the export of such wastes to the prohibiting parties; b) parties are to prohibit the export of hazardous wastes if the state of import does not consent in writing, in the case where the state of import has not prohibited the import of such wastes; c) parties are to prohibit all persons under their national jurisdiction from transporting or disposing hazardous or other wastes unless such persons are authorised to perform such types of operations; d) states of export shall not allow the generator of hazardous or other wastes to commence the trans-boundary movement until they have received written confirmation that the notifier has received the written consent of the state of import.

The Bamako Convention was signed by twenty-six African countries in 1991. Its objective was to create a framework of obligations to strictly regulate the trans-boundary movement of hazardous waste to and within Africa (UNEP 1996). The convention is confined to hazardous wastes, defined as substances banned, cancelled or refused registration by government regulatory action for health or environmental reasons. Radioactive and other wastes are listed in the annexes, but exclude wastes from ship discharges, which are covered by another convention. Fortunately, the convention provides some latitude by giving signatories the responsibility to enact legislation covering the identification and categorisation of hazardous wastes not listed in it. Specifically, it states as requirements: a) the exchange of information amongst signatory states on incidents of hazardous wastes, and on approaches to solutions to identified problems; b) the establishment of monitoring and regulatory authorities to report and act on trans-boundary movement of hazardous wastes; and c) cooperation between signatory states and with international organisations in the fulfilment of the objectives of the convention.

Implementation Constraints and Achievements for Africa

There are many international treaties aimed at protecting the environment. Therefore, a drastic reduction in the number and intensity of environmental problems might reasonably be expected. This, unfortunately, is not the case. Why? Africa's greatest priority and preoccupation is economic development. The challenge is to find policies that can enhance economic growth, whilst at the same time preserving the natural resource base. Formulating sustainable environmental policies requires an appropriate mix of economic incentives and suitable institutional arrangements, based on a clear specification of property rights (ECA 2002).

From the general economic situation in Africa, it could easily be concluded that economic incentives are a far-flung reality for a majority of the countries. Even the funding from international donor agencies could hardly be sufficient, let alone be allowed, to defray such costs. Institutional arrangements for the definition of property rights are in principle possible, but it still remains doubtful as to whether such policies can be implemented, especially in situations where governments are rapidly becoming aware that huge revenues that would have previously fed central treasuries are to be directed at communities.

Indeed, efforts to address environmental concerns pose numerous challenges to the African continent, among which are the following identified by ECA:

- the mobilization of the scientific community to mount an integrated programme for methods, standards, data collection, and research networks for assessment and monitoring of soil, water, land, forest and atmospheric degradation

- the development of environment use models that incorporate both natural and human-induced factors that contribute to degradation and that could be used for resource use planning and management

- the development of information systems that link environmental monitoring, accounting, and impact assessment to degradation

- the implementation of policies that encourage sustainable environmental resource use and management and assist in the greater use of environmental resource information for sustainable livelihoods

- the implementation of economic instruments for the assessment of environmental degradation and encourage the sustainable use of environmental resources' (ECA 2002:xii).

Cunningham et al. (2003) note that many treaties present, more-or–less, vaguely good intentions. This is due particularly to the fact that environmental protection cannot generally be legislated for, nor enforced. We may doubt the credibility of such a statement, considering in particular the numerous activities of the UN. But it should be clarified that the role of the UN and other regional organisations is to bring stakeholders together to negotiate solutions to common environmental problems. The key word here is 'negotiate'. It may be observed that the implementation of most negotiated solutions depends more on moral persuasion and public embarrassment than on actual enforcement. It is of no surprise that although many African countries are signatories to a number of international conventions relating to environmental resources management, many are still not committing sufficient resources to tackle the problem. Degradation related to global concerns, such as climate change, is simply not a priority for many governments regionally, though its potential importance is recognised globally. The truth is that there are almost no domestic or external pressures at present for African countries to implement policies related to global environmental problems, given the low level of greenhouse gas emissions in many of these countries and the possibility that there is a net sink for carbon dioxide on the continent (ECA 2002).

Another constraint is the general rule that international agreements are only binding on signatory parties (Shaw 1997). Many nations are unwilling to give up their sovereignty through the signing of such agreements (Cunningham et al. 2003). This means that while some nations might be struggling to address, say, the issue of hazardous waste dumping, others will continue to promote the problem, for economic gains. One of the principal problems of international conventions is the tradition of unanimous consent (Cunningham et al. 2003). A single recalcitrant nation has the power to veto the wishes of the vast majority. A case in point is the situation where even though more than 100 nations at the UN Conference in Rio de Janeiro in 1992 vouched for restrictions on the release of greenhouse gases, the US negotiators influenced the rewording of the convention. The changes meant that the convention 'urged', but did not 'require', nations to stabilise their emissions, with the implication that nations were not obliged to do so.

One fundamental challenge of environmental governance at a national level is the coordination of the multiple focal points of the various environmental agreements. Some African countries, such as Ghana, Uganda and Kenya have created national environmental management, or protection, agencies in a bid to overcome this problem (Menang 2007).

Overall, despite the numerous hurdles to their implementation, international environmental agreements have registered some great achievements for the African continent. These can be seen in two broad areas: socio-political reforms, and increasing environmental commitment. Socio-politically, a number of African states are beginning to adopt more devolved and inclusive ways of managing societies and their resources. The application of such obvious democratic principles finds expression not only in new environmental, forestry and wildlife laws, but also in founding constitutional, land and local government laws, as in the cases of South Africa, Lesotho, Namibia, Swaziland, Mozambique, Malawi, Zambia, Uganda, Tanzania, Kenya, Ethiopia, The Gambia, Burkina Faso, Mali, Senegal and Benin. At least thirty-five countries have enacted new codes in their forestry laws since 1990, or had these in draft by early 2002.

From the point of view of environmental commitment, a number of African governments, NGOs and some individuals have taken commendable initiatives to address environmental problems with varying degrees of success. UNCCD (cited in Minang 2007) reports on improvements in efforts to combat desertification in Burundi, Cape Verde, Djibouti, Ghana, Kenya, Morocco, Niger, Swaziland, Tunisia and Zambia. Some countries (Kenya and Burundi for example) have taken exemplary initiatives in the fight against desertification. In Kenya, the government created a national Environmental Management Authority in 1999. This semi-autonomous coordinating body for all multilateral environmental agreements draws members from all the major ministries, NGOs and research and academic institutions. The body promoted the creation of environmental committees at provincial and district levels to coordinate action on various environmental themes. An important element of the reforms in Kenya was the creation of the Desertification Community Trust Fund, launched in 2004. This is a government-funded programme that has supported projects in a number of districts. In Burundi, the land and forestry codes were revised to make them more compatible with actions to combat desertification. Measures were taken to serve as incentives for individuals and communities to invest time and resources in land rehabilitation activities. For example, land ownership rights were enhanced, especially for vulnerable groups such as widows and orphans, making it easier for them to obtain certificates that serve as deeds. Communities were encouraged to manage woodlands on communal land (Minang 2007). It is important to note that several other African countries have made significant progress in the area of community forestry, which will be elaborated in Chapter Four.

Economic instruments to effect environmental policy, particularly the application of the polluter-pays principle, emphasised in both the Rio Declaration and Agenda 21, are already being implemented by some African countries. For example, the Zimbabwean government introduced a carbon tax on all vehicles in its 2001 fiscal year. The levy charged depends on the engine capacity of the vehicle. In South Africa, the National Water Act of South Africa makes provision for the payment of pollution and catchment levies. These funds are specifically intended to promote the protection of water resources. They can be tapped to support both the capital and operational costs of providing a higher level of service. Other economic instruments for environmental policy applied in Africa take the form of tax exemptions and tax credits. For example, Zambia has reduced or abolished import duties on pollution control equipment and environmentally sound technologies. Similarly, Mauritius provides manufacturing enterprises with duty exemptions, tax credits, and other incentives around the importation of pollution control equipment and environmental protection to facilitate economic, industrial and technological development (UNEP cited in ECA 2002).

Conclusion

This chapter has attempted a definition of international law, highlighting its development and environmental application. It focuses on international environmental law and its implication in a world beset by numerous environmental problems, which have both local and global impacts. We have noted that many efforts have been made to address these problems, at national, regional and international levels, and that problems with regional or global impacts can best be addressed through international cooperation.

It is clear that international environmental law has played an increasingly important role in addressing environmental problems, with regional and global implications. This has resulted in a proliferation of international treaties or conventions covering a wide range of environmental concerns. Of over 200 such conventions, just eight have been addressed: biodiversity, wetlands of scientific importance, endangered species, desertification, ozone layer depletion, climate change, and hazardous wastes. These include issues that may appear remote, but are significant to Africa, such as ozone layer depletion and climate change; and those that are immediately understood as constituting a problem on the continent, such as desertification, endangered species and hazardous waste dumping.

Despite their role in addressing environmental problems, international conventions are not without implementation constraints. First, there is an absence of a body that can enforce environmental protection at an international level. This absence is felt to the extent that negotiated solutions to global environmental problems rely more on moral persuasion and public embarrassment. The second factor is that international conventions are binding only on signatory nations, meaning that those refusing to sign are free to continue perpetrating environmental problems. Finally, the tradition of unanimous consent makes it possible for a single nation to weaken the strength of international conventions.

In spite of the numerous constraints, international agreements have brought about some important reforms in Africa, with practical action on the ground. These achievements can be seen in the improving socio-political arrangements and growing environmental commitment in a number of countries.

Revision Questions

1. What is Environmental Law?

2. List and discuss the two sources of international environmental law, highlighting the strengths and weakness of each.

3. Why is the application of international environmental law necessary in environmental protection?

4. What are treaties? Discuss the stages required for them to come into force.

5. Name and discuss three international environmental agreements, highlighting when they were signed and their objectives.

6. Discuss the constraints in the implementation of international conventions in Africa.

7. What are the positive impacts of the international agreements on the African continent?

Critical Thinking Questions

1. Climate change threatens global economies but offers development opportunities for developing countries. Discuss.

2. Why is the Clean Development Mechanism of the Kyoto Protocol so called, and how should Africa engage with the mechanism?

3. What are the opportunities and constraints of Africa becoming one of the actors in the implementation of the Kyoto Protocol?

4. Despite the Basel and Bamako conventions, waste dumping has continued to constitute a serious problem, especially for Africa. What are the factors promoting this unfortunate practice?

References

Agrawala, S., Gigli, S., Raksakulthai V., et al., 2005, 'Climate Change and Natural Resource Management: Key Themes from Case Studies', in S. Agrawal, *Bridge over Troubled Waters: Linking Climate Change and Development*, Paris: OECD, pp. 85–132.

Baumert, K. and Pershing, J., 2004, *Climate Data: Insights and Observations*, Arlington: Pew Center on Global Climate Change.

Cunningham, W.P., Saigo, B.W., and Cunningham, M.A., 2003, *Environmental Science: A Global Concern*, New York: McGraw-Hill.

ECA, 2002, *Economic Impact of Environmental Degradation in Southern Africa*, Lusaka: Economic Commission for Africa.

French, H.F., 1992, *After the Earth Summit: The Future of Environmental Governance*, Worldwatch Paper No. 107, Washington DC: Worldwatch Institute.

Glahn, G., 1970, *Law among Nations*, Toronto: The Macmillam Company.

Hugh, D., 1992, *Environmental Law*, London: Butterworths.

Indriani, G., 2005, *Gas Flaring Reduction in the Indonesian Oil and Gas Sector – Technical and Economic Potential of Clean Development Mechanism (CDM) Projects*, Hamburg: HWWA Report 253.

Manguiat, M.S.Z., Verheyen, R., Mackensen J., and Scholz, G., 2005, *Legal Aspects in the Implementation of CDM Forestry Projects*, Gland, Switzerland and Cambridge, UK: IUCN.

Minang, P.A., 2007, *Implementing Global Environmental Policy at Local Level: Community Carbon Forestry Perspectives in Cameroon*, a PhD dissertation submitted to the University of Twente.

Ramsar Secretariat, 2007a, *Ramsar Information Paper*, No. 2, Ramsar Secretariat.

Ramsar Secretariat, 2007b, *Ramsar Information Paper*, No. 4, Ramsar Secretariat.

Shaw, M.N., 1997, 4th edition, *International Law*, Cambridge: Cambridge University Press.

Shimada, K., 2004, 'The Legacy of the Kyoto Protocol: Its Role as the Rulebook for an International Climate Framework', *International Review for Environmental Strategies*, Vol. 5, No, 1, pp. 3–14, Kamiyamaguchi: Institute for Global Environmental Strategies.

Umozurike, U.O., 1995, *Introduction to International Law*, Ibadan: Spectrum.

UNEP, 1996, *Register of International Treaties and Other Agreements in the Field of the Environment*, Nairobi: UNEP.

UNEP, 2000, *The Montreal Protocol on Substances that Deplete the Ozone Layer*, Nairobi: UNEP, <http://www.unep.org/ozone>

UNEP, 2003, *Convention on Biological Diversity*, Montreal: UNEP/CITES,<www.unccd.entico.com/english/faq.htm>

Chapter 4

Environmental Management on African Agenda

Introduction

Insofar as global landmarks for environmental management have sought to influence environmental policy at an abstract level, their impact has been great in influencing continental and sub-continental environmental policies. However, Africa's environmental management agenda still remain very much in the making of donors and other development agencies, such as the World Bank, SIDA, CIDA and DANIDA, which hold financial resources, and the capacity to influence political events. It was only recently that key African institutions such as the Africa Union and the African Development Bank Group have come up with policies governing environmental management. The continent has also engaged in a number of controversial multilateral environmental conventions, such as the Bamako and Basel Conventions on hazardous waste management, and the Kyoto Protocol.

Besides discussing aspects pertaining to the donor community and its environment agenda for Africa, in this chapter the following policy frameworks are discussed: the Bamako Convention of 1991 (OAU Secretariat 1991), the New Partnership for Africa's Development (NEPAD 2002), Southern African Development Community's Policy and Strategy for Environment and Sustainable Development (SADC 1996), and the 2003 Regional Indicative Strategic Development Programme (SADC 2003), with the aim of establishing which environmental agenda are being followed.

We cannot underestimate (nor should we overemphasise) the role that the United Nations Environment Programme (UNEP) has played in shaping Africa's environmental agenda. This is mainly due to its close proximity to the African continent, being the sole UN agency with its headquarters in Africa, in Nairobi.

Environmental Concerns in Africa

Environmental concerns in Africa can be mapped as a web, as presented in Figure 4.1. Some of these concerns will be discussed indepth in Chapter Five. Although most of the concerns highlighted in Figure 4.1 are restricted to the African continent, others such as the trans-boundary movement of hazardous chemicals, climate change, and global warming are of an international nature. It is critical to understand these environmental concerns as a web, to be addressed in a holistic manner.

Most donors have failed to do so over the past few decades, probably due to limits in financial resources and local human capacity.

Figure 4.1: Summary of Key Environmental Concerns in Africa

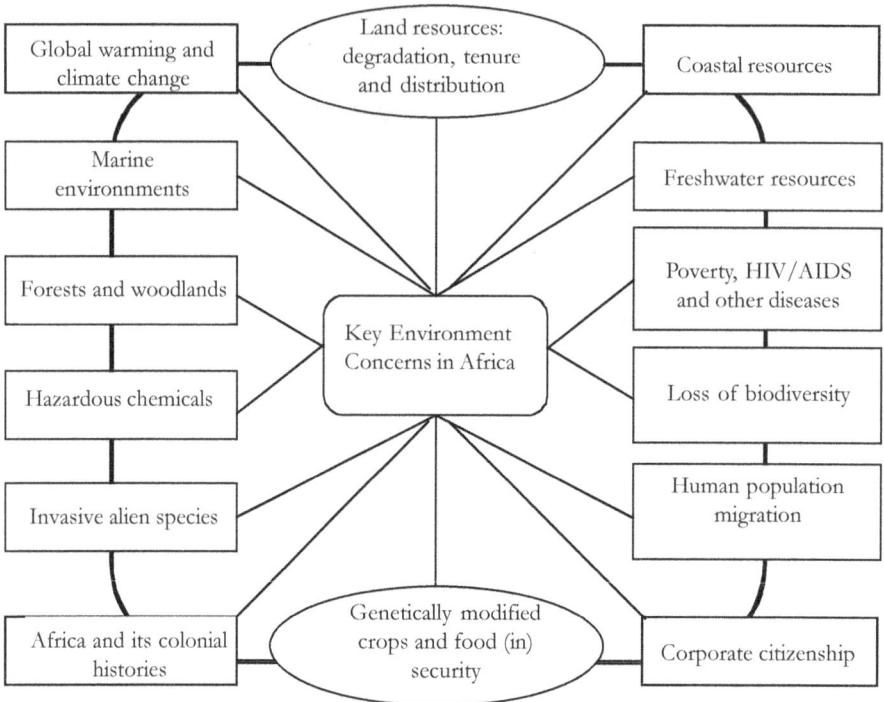

Although Africa is faced with a multitude of environmental concerns, climate change has been identified as a critical environmental and sustainability concern for the twenty-first century. Key environmental issues and crises are increasingly linked to climate change. The changing climate impacts on phenomena such as land resources, freshwater, marine environments, forests and woodlands, biodiversity, poverty, invasive alien species, use of chemicals, disease, food security and corporate citizenship.

UNEP's Role in Shaping Africa's Environmental Agenda

UNEP has been instrumental in shaping both the international and African environment agenda. Its mission is stated thus: 'To provide leadership and encourage partnership in caring for the environment by inspiring, informing, and enabling nations and peoples to improve their quality of life without compromising that of future generations'. To this end, a number of milestones in setting and refining the environmental agenda have been witnessed, as indicated below.

• 1972 - UN Conference on the Human Environment recommends creation of a UN environmental organisation

- 1972 - UNEP created by UN General Assembly
- 1973 - Convention on International Trade in Endangered Species (CITES)
- 1975 - Mediterranean Action Plan first UNEP-brokered Regional Seas agreement
- 1979 - Bonn Convention on Migratory Species
- 1985 - Vienna Convention for the Protection of the Ozone Layer
- 1987 - Montreal Protocol on Substances that Deplete the Ozone Layer
- 1988 - Intergovernmental Panel on Climate Change (IPCC)
- 1989 - Basel Convention on the Transboundary Movement of Hazardous Wastes
- 1992 - UN Conference on Environment and Development (Earth Summit) publishes Agenda 21, a blueprint for sustainable development
- 1992 - Convention on Biological Diversity
- 1995 - Global Programme of Action (GPA) launched to protect marine environment from land-based sources of pollution
- 1997 - Nairobi Declaration redefines and strengthens UNEP's role and mandate
- 1998 - Rotterdam Convention on Prior Informed Consent
- 2000 - Cartagena Protocol on Biosafety adopted to address issue of genetically modified organisms
- 2000 - Malmö Declaration - first Global Ministerial Forum on the Environment calls for strengthened international environmental governance
- 2000 - Millennium Declaration - environmental sustainability included as one of eight Millennium Development Goals
- 2001 - Stockholm Convention on Persistent Organic Pollutants (POPs)
- 2002 - World Summit on Sustainable Development
- 2004 - Bali Strategic Plan for Technology Support and Capacity Building
- 2005 - Millennium Ecosystem Assessment highlights the importance of ecosystems to human well-being, and the extent of ecosystem decline
- 2005 - World Summit outcome document highlights key role of environment in sustainable development.

Donors and the African Environmental Agenda

Africa's environmental agenda has been, and is currently, strongly driven by donors and aid agencies. However, more recently, African institutions such as the African Development Bank, have also established environmental agenda for the continent.

The World Bank's Environment Strategy

The World Bank's (hereafter referred to as the Bank) environmental agenda was partially shaped by criticisms of its pro-economic developmental agenda from the mid-1980s to mid-1990s. To this end, the Bank sought to broaden its environment agenda, particularly in the late 1990s (World Bank 1998), by accepting that sustainable development can only be achieved if activities are sustainable at both global and local levels. The Bank decided that helping partner countries to improve their environmental management capacity and mainstreaming environmental sustainability principles into development programmes were at the core of its business. Pilot projects to test new instruments for environmental management, such as the Ugandan Institutional Capacity Project for Wildlife and Tourism, focusing on strengthening institutions, staff training and public/private partnerships to promote ecotourism, were funded. Pollution abatement projects were also established in Egypt in 1998. In South Africa, 'green accounting' initiatives were launched in the same year.

The Bank's ultimate environment agenda was set out in the document: *Making Sustainable Commitments: An Environment Strategy for the World Bank* (World Bank 2001). The environment strategy outlines how the Bank intended to work with client countries to address their environmental challenges and ensure that Bank projects and programmes integrate principles of environmental sustainability.

Prior to 2000, one of the central pillars, in terms of environmental concerns, for the Bank, had been the development of National Environmental Action Plans (NEAPs). Many of the Bank's client countries were assisted in developing NEAPs. The main criticism was that they were not adequately implemented. They failed to network into other ongoing government initiatives aimed at addressing environmental concerns, particularly on a national scale.

In its 2006 annual report (World Bank 2006), the Bank recognised the need to work towards the achievement of the MDGs, including Goal 7, which aims at ensuring environmental sustainability. The Bank focuses on supporting agriculture in Africa. Agriculture is linked to integrated natural resource management such as protecting watersheds, sustaining increased investment in irrigation and drainage, supporting effective forest law enforcement and sound governance, and promoting efforts to enhance the resilience of farms, forests, and fisheries to climate change. The Bank is also involved in a new multi-donor initiative, code-named PROFISH, addressing the threat of collapsing fish stocks. Furthermore, the global threat of *zoonoses* (diseases that can be transmitted from animals to humans), such as 'mad cow' disease and avian flu, has brought the issue of livestock management and animal health back to the top of the development agenda, for the World Bank, and globally.

The Bank has also stated that global warming and climate change are at the top of its developmental agenda. The Bank reckons that the international community is faced with a serious crisis of securing affordable and cost-effective energy supplies. In this regard, it has put in place a Clean Energy and Development Strategy aimed at addressing the energy needs of developing and newly industrialised countries. The strategy involves addressing bottlenecks to accessing energy services, controlling greenhouse gas emissions, and helping developing countries adapt to climate risks (World Bank 2006). In the eyes of the Bank, an extensive array of technologies exist that can provide the requisite energy services. But policy reform in the energy sector is urgently needed to stimulate roughly US$300 billion of investment required per year. With regard to climate change, the Bank acknowledges that the poorest countries, especially in Africa, and their poorest citizens are among the most vulnerable. As such, adapting to climate variability and change is a priority for developing countries and will require the transfer of existing technologies, the development of new ones, and the revision of existing planning standards and systems.

In supporting environmental sustainability, the Bank assists developing countries in meeting the cross-cutting goal of ensuring environmental sustainability by integrating environmental concerns into all development-related work (World Bank 2006). This is done through the implementation of its environment strategy (World Bank 2001). It addresses the links between environment, poverty and economic growth, with a particular emphasis on the health, livelihoods and the vulnerability of poor people. The Bank is also working with partner countries to systematically evaluate their environmental priorities, the environmental implications of key policies, and their capacity to address development priorities and related environmental concerns. To monitor progress towards achieving the MDGs, focusing on environmental sustainability, the Bank has put in place a set of measures to assess changes in natural wealth, which are taking place in developing countries. A summary of the Bank's lending to borrowers in Africa from 2001–6 is shown in Table 4.1.

Table 4.1: World Bank Lending to Borrowers in Africa (US$, millions)

THEME	2001	2002	2003	2004	2005	2006
Economic Management	138.5	138.7	37.8	68.0	46.5	31.4
Environmental and Natural Resource Management	110.0	159.9	227.0	195.2	217.2	250.6
Financial and Private Sector Development	625.8	780.7	383.6	810.9	768.2	979.1
Human Development	399.4	739.0	811.4	618.2	620.2	673.3
Public Sector Governance	429.6	851.9	432.4	818.4	708.0	964.7
Rule of Law	34.0	22.5	34.5	28.3	30.9	179.7
Rural Development	296.3	329.2	384.1	360.7	537.2	528.6
Social Development, Gender, and Inclusion	491.8	347.4	420.0	374.3	221.8	198.5
Social Protection and Risk Management	376.4	98.3	543.7	209.2	294.3	262.7
Trade and Integration	261.5	46.4	37.2	371.5	232.0	413.1
Urban Development	206.1	279.6	425.5	261.1	211.4	304.9
Theme Total	3,369.6	3,793.5	3,737.2	4,115.9	3,887.5	4,786.6

Source: World Bank, 2006:33.

It is of interest to note how the Bank has gradually increased its lending towards the environmental and natural resource management sector. The amount allocated has more than doubled since 2001. This explains how the Bank is actively involved in shaping environmental agenda on the continent. The bottom line is that only environmental and natural resource management projects, in which the Bank has an interest and are eligible for the Bank's borrowing benefit can receive funding. This alone creates an uneven terrain in addressing environmental concerns on the continent.

CIDA's Policy for Environmental Sustainability

In 1992, CIDA put in place its policy for environmental sustainability. The policy acknowledges the global nature of a number of environmental problems. The conditions necessary to sustain development and quality of life in developed and developing nations are linked. The degradation of natural resources is identified as the key factor limiting development potential of countries in the South and the North alike (CIDA 1992). The policy document identifies factors limiting developing nations to address both local and global environmental problems. Such factors include poverty, and the inequitable control of resources, coupled with differing attitudes and levels of knowledge about the environment on the part of both individuals and societies. Other factors outlined are macroeconomic pressures, such as low commodity prices and debt burdens, inadequate financial and human resources, underdeveloped institutional and technological capabilities, inadequate opportunities for people to participate meaningfully in the development process, inappropriate or narrow economic and social policies, and inadequate incentives for environmentally-sound behaviour.

CIDA's environmental agenda in Africa and other developing regions dates back to four decades. In the early 1980s, CIDA supported the creation of environmental monitoring and teaching institutions. The agency also adopted an environment policy that led to the application of procedures for routine environmental screening and assessment of development projects in the late 1980s. These initiatives have resulted in CIDA being recognised as one of the key stakeholders in environmental management on the continent. The agency has also provided support to NGOs, universities and private-sector organisations striving to integrate environmental considerations into development initiatives. Many of CIDA's country and response programmes have resulted in the preparation of environmental strategies as part of the programme plans. In the case of Zimbabwe, CIDA was actively involved in the development of the National Environmental Management Act that passed through parliament in December 2002. Regardless of its – much appreciated – efforts in assisting developing countries to address environmental damage, CIDA believes that much is outstanding. To this end, its 1992 Policy for Environmental Sustainability recognises that there is need for an interdisciplinary approach in analysing programme and project design, as well as following cross-sectoral ecosystem approaches to project implementation. The agency also advocates the promotion of sound domestic and international economic policies. The 1992 policy acknowledges that environmental problems are often severe in poor communities. They tend to arm the disadvantaged and disempowered disproportionately: the poor, the disabled, women, children and indigenous peoples. The policy calls for renewed efforts, both from the rural and urban communities, to address the socio-economic needs of such disadvantaged groups, an aspect which would arguably yield environmental benefits. This policy for environmental sustainability stresses its mission as supporting sustainable development in developing countries. Sustainable development is taken as embracing the five pillars of economic, environmental, social, cultural and political sustainability. Hence, there are five related aspects to the concept of sustainability (CIDA 1992:5–6).

Achieving Economic Sustainability

This requires appropriate economic policies, efficient resource allocation and use, more equitable control over resources, and increased productive capacity among the poor.

Achieving Social Sustainability

This means more equitable income distribution, and ensuring the participation of intended beneficiaries, and those who may be affected, in the decision making which affect their lives.

Attaining Cultural Sustainability

This requires sensitivity to cultural factors, including cultural diversity, and a recognition of cultural values conducive to development.

Attaining Political Sustainability

This aspect is premised on the assurance of human rights and the promotion of democratic development and good governance.

Achieving Environmental Sustainability

Ecosystems should be protected and managed to maintain both their economically productive and their ecological functions, the diversity of life in both human-managed and natural systems, to protect the environment from pollution, and maintain the quality of land, air and water.

In line with the need to help developing countries achieve environmental sustainability, CIDA's policy for environmental sustainability seeks to pursue the following environmental sustainability and operational objectives.

Environmental Sustainability Objectives

- increase the institutional, human resource and technological capacities of developing country governments, organisations and communities to plan and implement development policies, programmes and activities that are environmentally sustainable;

- strengthen the capability of developing countries to contribute to the resolution of global and regional environmental problems, while meeting their development objectives.

Operational Objectives

- ensure that environmental considerations, including opportunities for enhancing environmental sustainability, are integrated into sector and cross-sector programmes, programme assistance, and project planning and implementation, taking into account views of beneficiaries and local communities;
- promote and support environmental and broader socio-economic policy dialogue, programme assistance and projects that directly address environmental issues;
- implement design measures that minimise negative environmental impacts and enhance environmental benefits of projects, or identify alternatives;
- encourage and support Canadian, international and developing country partner organisations to develop policies, programmes and projects that further the objectives of environmental sustainability;
- contribute to the development of knowledge and experience in Canada and in developing countries, on undertaking environmentally sustainable forms of development;
- promote education and awareness among governments and the public in Canada and in developing countries of the importance of environmentally sustainable approaches to development.

That environmental sustainability encompasses political aspects, strengthens our earlier arguments that the environment is as much a political matter as an environmental and social matter. However, CIDA's policy is still rather silent on how it plans to address sensitive enviro-political issues, notably the land question in Africa.

DANIDA's Environmental Strategy, 2004–2008

The Strategy for Denmark's Environmental Assistance to Developing Countries 2004–2008 outlines targets, principles and priorities for Denmark's environmental and environmentally related assistance to developing countries. The strategy recognises poverty alleviation as the outstanding challenge and as a prerequisite for stable, sustainable development (DANIDA 2004). The Danish government also realises that it is necessary to enhance the environmental sustainability of aid programmes, so as to secure global stability and development. To this end, the development policy was strengthened through a significant focus on the environment. Again, we witness the environmental management agenda being widened to capture poverty, to which early agenda found it difficult to relate, and integrate into environmental management. According to DANIDA, the strengthening of environmental assistance in developing countries involves the incorporation of special environmental assistance into general development programmes.

The environment, as a cross-cutting issue, will receive increased attention, with international environmental activities allocated increased funding. One of the key environmental problems on the Danish agenda is the effort to combat global warming. This includes a focus on the correlation between environmental assistance and the use of the Kyoto Protocol's flexible mechanisms, particularly the CDM. Hence over time, and in view of the dynamism of environmental problems and challenges, DANIDA is spearheading an agenda to address global warming.

The DANIDA strategy stands out as the basis for the actual implementation of environmental assistance in developing countries. It covers an entire range of activities in developmental cooperation, including: environment as a cross-cutting issue, environment in multilateral cooperation, environmental issues in programme cooperation, and special environmental assistance. The integration of special environmental assistance into regular development sector programmes implies that the aim of environmental assistance to developing countries is to increase sustainable development, and as an integrated part of poverty alleviation to limit the damage to the environment. Developing countries will be assisted with environmental challenges by enhancing their ability to take responsibility for the environment themselves.

The strategy will be implemented through partnerships with the cooperating countries. It is based on national strategies for poverty reduction and sustainable development. Other issues of interest include gender equality, human rights and democracy (including good governance). These are considered cross-cutting issues. They also feature on CIDA's agenda for environmental management. Likewise, multilateral environmental efforts will be governed by the principles of active multilateralism. Coordination and synergy between bilateral and multilateral environmental efforts will be a priority, for example, in the implementation of international conventions, and in the attention paid to the environment as a cross-cutting issue in the bilateral cooperation components of the work of international organisations. Environmental assistance is governed by existing guidelines for development aid, including 'A Strategy for the Support to the Development of Civil Society' (which also covers cooperation with Danish NGOs) and 'Aid Management Guidelines'. Special development assistance covers poor countries as well as a number of middle-income ones. The existing assistance has contributed to the development and implementation of the protection of nature and the environment, and to sustainable development. In middle-income countries, with rapidly growing environmental problems, assistance has contributed to efficient solutions that have also benefited other countries in the respective regions. Special environmental assistance will therefore be sustained in poor and middle-income countries.

Environmental assistance will simultaneously support the implementation of the Danish government's climate strategy, through the development of projects that can subsequently lead to purchases of carbon credits from developing countries.

The contribution towards the fulfilment of the Millennium Development Goals (MDGs) is another benchmark against which Danish environmental assistance will be measured. The DANIDA strategy recognises that only a few countries have prioritised the relationship between poverty and the environment in their national poverty reduction strategies. This element cannot be over-emphasised, bearing in mind that the World Bank and International Monitory Fund mechanisms, under which most African countries' poverty reduction strategies were developed, did not have an environmental management component. The DANIDA strategy calls for a need to develop methods and tools to ensure this happens more frequently. The strategy also advocates the need to continue efforts to make the environment a priority in international organisations, and in development banks. Probably, this is the reason why the African Development Bank Group is initiating a policy for addressing Africa's environmental challenges. Special environmental assistance has had particularly good results in projects involving urban environment, sustainable energy, and in the management of natural resources. South Africa, for example, has embarked on a bilateral Urban Environmental Management Programme funded by DANIDA, running from 2006 to 2010.

The main priorities and operational targets for Denmark's international environmental assistance and cooperation with the developing countries for the period 2004–8 are to (DANIDA 2004:7–8):

- improve sustainable development and limit environmental degradation at global, national and local levels within the overall development policy objective of poverty alleviation. This will happen through assistance in management of environmental challenges and by enhancing the developing countries' ability ultimately to bear the responsibility themselves;

- upgrade the environment as a crosscutting issue in the bilateral as well as multilateral development cooperation. Special attention will be paid to the incorporation of environmental issues in national poverty reduction strategies, to national strategies for sustainable development and to environmental analyses in sector programmes;

- continue to prioritise assistance to focused development and the implementation of international environmental conventions and agreements – including the plan of action from the World Summit for Sustainable Development that took place in 2002 in Johannesburg – in bilateral and multilateral development cooperation;

- upgrade and strengthen international environmental cooperation within multilateral assistance – particularly with regard to efforts involving water, energy, chemicals and climate. This will include increased contributions to the global climate foundations;

- work for a substantial replenishment of funds in the Global Environmental Facility (GEF);

- focus the bilateral and the special environmental assistance on three areas: urban and industrial environment, sustainable energy and management of natural resources;

- continue special environmental assistance in Southern Africa, in the Republic of South Africa, Mozambique, Zambia and Tanzania and to start a similar programme in Kenya;

- re-organise the special environmental assistance by applying a long-term programme approach in line with the sector programme assistance used in traditional development aid;

- concentrate cooperation on Clean Development Mechanism (CDM) projects in five countries: South Africa, Thailand, Malaysia, Indonesia and China.

These points set a clear environmental management agenda for Africa. In addition, the specification of countries with which DANIDA will be working in the context of its special environmental assistance programme brings an interesting dimension to donor aid. For example, in SADC, DANIDA will be working with South Africa, Mozambique and Tanzania. In East Africa, it will be working with Kenya. DANIDA will also be concentrating on cooperation over the CDM projects with South Africa. Hence, the CDM environmental agenda is being promulgated, but with a single country on the continent: South Africa.

African Development Bank Group's Policy on the Environment

The 2004 African Development Bank Group's Policy on the Environment (hereby referred to as the policy) was driven by a number of factors: the recognition and acceptance of sustainable development as the dominant development paradigm for this century; a need for a greater focus on pro-poor growth policies and programmes to counter unacceptable impoverishment rates; rapid progress in the inevitable integration of Africa into the globalisation process; and the need for improved governance with a clearer commitment by the majority of African governments to provide the necessary leadership for sustainable development (African Development Bank 2004:iii).

The Bank's policy acknowledges the significant progress made in the implementation of Agenda 21, adopted at the 1992 Rio Earth Summit; the ratification of a large number of environmental conventions, agreements and protocols; and the growing use of MDGs as a measure of development. The policy recognises the ongoing degradation of the environment across the continent, in spite of significant strides made at national and regional levels to establish the necessary legal and institutional frameworks. But, above all, the policy takes into consideration the various opportunities, in terms of resources and skills, available to Africa for its development, and for the improvement of the overall quality of life of its people. Critical is its acknowledgment of the availability of African capacity to deal with its own environmental challenges.

The Bank's policy framework embraces the concept of sustainable development as expressed in the provisions of the 1987 Brundtland Commission. It recognises, however, that although the principles of sustainability have been globally accepted for decades, their translation into specific environmental management strategies has been fraught with practical and theoretical problems. However, the growing evidence of the rapid deterioration of ecological capital and diminishing assimilative capacities of ecosystems, coupled with the global scale of environmental problems, have forced policymakers to rethink their development strategies, and to accept that the environment and the economy are interdependent. Sustainable development is widely recognised as the preferred development paradigm.The environment policy stresses the anticipatory nature of sustainable development, rather than the reactive responses so predominant in development-related decisions. In adopting the concept of sustainable development as an environmental policy framework, care is taken to recognise the considerable constraints on Africa.

The Bank acknowledges that large parts of the continent are still threatened by growing poverty and disparity in wealth distribution, both nationally and regionally. Faced with such constraints, many African countries have not been able to make progress in the transition from normative standards to operational programmes in sustainable development. The Bank also attests to the effects of globalisation, and the decreasing flows of development assistance, which have significantly and negatively affected the ability of the poorest nations to deal with environmental challenges. However, the policy recognises that there have been a number of achievements: awareness creation, capacity building, reinforcement of legislative, institutional and regulatory frameworks, and integration of environmental concerns into national economic development strategies. New threats have, however, surfaced since 1992, with HIV/AIDS becoming a major development crisis. Sub-Saharan Africa alone represents 70 per cent of the world's population living with HIV/AIDS.

The Bank's policy acknowledges that Africa is endowed with a rich resource base consisting of minerals, rich flora and fauna, and large tracts of rainforests. These can provide a basis for industrial development and eco-tourism. Its rich biodiversity constitutes essential global public goods that not only provide sources of non-timber forest products, but also help attract larger amounts of international funding for preservation. Economic growth, however, translates into increased economic activities that generate social and environmental costs and benefits. Such externalities constitute potent threats to the economic system, leading to market failures. To this end, there is a need to enhance the positive, whilst integrating social and environmental concerns into economic development policies to reduce or internalise the negative externalities.

From the Bank's viewpoint, environmental concerns are now routinely integrated into country dialogue processes and project design. There is now widespread recognition among Bank and regional member country staff that sustainable development and poverty reduction cannot be divorced from the global environment, and that local-to-global linkages must be made. The Bank also recognises the importance of cooperation with sub-regional and international partners in environment policies to build on efforts aimed at enhancing environmental management.

In line with the conceptual framework adopted to handle the development challenges and priorities facing Africa, the policy supports the creation of the right conditions that allow as many stakeholders as possible to play their role in achieving sustainable development. The overall goals of the policy are two-fold: to help improve the quality of life of the people of Africa; and to help preserve and enhance the ecological capital and life-support systems across the continent. Some of the principles that have guided the development of the policy include the recognition that a strong and diversified economy constitutes a just means to enhance the capacity for environmental protection; environmental management tools, such as environmental assessments, should be used as a sustainability assurance, rather than as a mechanism of mitigation; community involvement, in particular of the most marginalised and vulnerable groups, in decisions that affect them, must be provided for; and governance structures and institutions most responsive to the needs and priorities of affected communities in general, and poor people and vulnerable groups in particular, should be encouraged.

The policy replaced the Bank's traditional sector-by-sector approach in the management of natural resources with cross-sectoral environmental policy actions based on an integrated approach. Such an approach ensures optimum results that will simultaneously help to meet basic human needs and protect the environment. The key environmental issues identified include: reversing land degradation and desertification, protecting the coastal zone, protecting global public goods, enhancing disaster management capabilities, promoting sustainable industry, increasing awareness, institutional and capacity building, environmental governance, urban development and population growth, and civil society organisations (African Development Bank 2004).

To assist in the implementation of the policy, the Bank is using a set of approaches that include: mainstreaming environmental sustainability considerations in all the Bank's operations; strengthening existing environmental assessment procedures and developing new environmental management tools; clearly demarcating internal responsibility in implementation; assisting in the building of adequate human and institutional capacity to deal with environmental management; improving public consultation and information disclosure mechanisms; building partnerships to address environmental issues, harmonise policies, and disseminate environmental information; and improving the monitoring and evaluation of operations. Specifically, the Environmental and Social Assessment Procedures released in 2001 will be fully enforced for all lending operations of the Bank. As policy lending is becoming increasingly important, Strategic Impact Assessment (SIA) procedures will be developed to bring environmental assessments upstream, by assessing the impacts of policies, programmes and plans, rather than conducting environmental impact assessment at project level.

New Partnership for Africa's Development

To address the ills of unsustainable development, NEPAD has put in place an Action Plan for the Environment Initiative (NEPAD 2002). The initiative notes that Africa is rich in natural resources, including land, minerals, biological diversity, wildlife, freshwater, fisheries and forests. However, rapid population growth, rising poverty levels (including the widening gap between the rich and the poor) and inappropriate development practices are major factors leading to degraded environments (UNEP 2003). The NEPAD initiative sets seven action plans, grouped according to areas of concern or programmes. Some of the programmes include: integrated waste and pollution control, the management of cities, and the management of coastal and marine resources (NEPAD 2002).

Sub-regional Environmental Sustainability Initiatives

At the SADC sub-regional level, issues of sustainability and the environment are dealt with in two documents: the 1996 SADC Policy and Strategy for Environment and Sustainable Development (SADC 1996), and the 2003 Regional Indicative Strategic Development Programme (SADC 2003). The Strategy for Environment and Sustainable Development was drawn up following the Rio Summit in 1992. The document provides an overall framework to guide good environment management. Five broad strategic areas that would facilitate sustainability in development were identified: assessing environmental conditions, trends and programmes made and needed for sustainable development; minimising significant threats to human beings, ecosystems and future developments; a call to move away from unsustainable to sustainable development for the benefit of all generations; managing shared natural resources in an equitable and sustainable manner; and increasing regional integration and capacity building for sustainable development.

The implementation plan was designed around sectoral responsibilities that were shared by the member countries. Some of the sectoral responsibilities included environment and land management, given to Lesotho, mining (Zambia), energy (Angola), fisheries (Malawi) and food security (Zimbabwe). In 2003, SADC launched its Regional Indicative Strategic Development Plan – RISDP (SADC 2003) aimed at re-orienting developmental strategies. RISDP re-grouped sectoral responsibilities under new cluster directorates that include trade, industry, finance and investment; infrastructure and services; food, agriculture and natural resources; and social and human development and special programmes. The environment is dealt with in the food, agriculture and natural resources cluster. RISDP takes the environment and sustainable development as one of eight cross-cutting priority areas. The document also recognises efforts made by member states to address environmental concerns since the early 1980s. These are reflected in the ratification of major multilateral environmental agreements such as the United Nations Framework Convention on Climate Change, Convention to Combat Desertification, Convention of Biological Diversity, Basel/Bamako Conventions, the Millennium Development Goal and many more. However, RISPD notes that there are still high levels of pollution as well as poor sanitation and urban conditions whereby the poor become both victims and agents of environmental degradation.

To address problems associated with environmental degradation (SADC 2003: 101), RISDP sets the overall goal of environmental intervention so as 'to ensure the equitable and sustainable use of the environment and natural resources for the benefit of present and future generations'. Five areas of focus are established: create the requisite harmonised policy environment as well as legal and regulatory frameworks; promote environmental mainstreaming in order to ensure the responsiveness of all SADC policies, strategies and programmes for sustainable development; ensure regular assessment, monitoring and reporting on environmental conditions and trends in the region; capacity building, information sharing and awareness raising on problems and perspectives in environmental stewardship; and ensure coordinated regional positions in the negotiations and implementation of multilateral environmental agreements.

Seven broad strategies are set out. Similarly, nine broad targets are set with instruments for regional cooperation. These were due for finalisation in December 2006. Environmental standards and guidelines developed were being implemented in 2008. State of the environmental reports for southern Africa are to be produced regularly at five-year intervals. SADC's plan of action for implementing the WSSD Implementation Plan was in place by 2004, and the principle of sustainable development should be integrated into national policies and programmes by 2015.

Controversy on Hazardous Waste Management

The Bamako Convention on the Ban of the Import into Africa and the Control of Transboundary Movement and Management of Hazardous Wastes within Africa provides guidelines for the management of hazardous wastes by member states of the OAU. It is named after the Malian capital, Bamako, in which it was adopted. The Bamako Convention was adopted on 30 January 1991 (OAU Secretariat 1991) as an alternative to the Basel Convention on the Control of Transboundary Movements of Hazardous Wastes (Basel Convention) adopted in March 1989 in Switzerland. OAU members felt that their concerns regarding the total ban on exporting hazardous wastes into Africa and other relevant regional aspects were not adequately addressed by the Basel convention. Article 1 provides several definitions, among them, for 'wastes', 'hazardous wastes' and 'transboundary movement'. Under Article 1(1), wastes are defined as 'substances or materials which are disposed of, or are intended to be disposed of, or are required to be disposed of by the provisions of national law'. Hazardous wastes, as per the provisions of Article 1(2) means wastes, as specified in Box 4.1, and those with characteristics outlined in Table 4.2. Under Article 1(4), trans-boundary movement refers to 'any movement of hazardous wastes from an area under the national jurisdiction of any State to or through an area under the national jurisdiction of another State, or to or through an area not under the national jurisdiction of another State, provided at least two States are involved in the movement'. Also of interest is what the Bamako Convention refers to as 'environmentally sound management of hazardous wastes', which in Article 1(10) means 'taking all practicable steps to ensure that hazardous wastes are managed in a manner which will protect human health and the environment against the adverse effects which may result from such wastes'. The Bamako Convention also stipulates definitions (Articles 11–23) in relation to 'area under the national jurisdiction of a State', 'state of export', 'state of import', 'state of transit', 'states concerned', 'person', 'exporter', 'importer', 'carrier', 'generator', 'disposer', 'illegal traffic', and 'dumping at sea' (OAU Secretariat 1991).

Box 4.1: Streams of Hazardous Wastes

The taxonomy of hazardous waste is as follows:

Y0 All wastes containing or contaminated by radionuclides, the concentration or properties of which result from human activity

Y1 Clinical wastes from medical care in hospitals, medical centers and clinics

Y2 Wastes from the production and preparation of pharmaceutical products

Y3 Waste pharmaceuticals, drugs and medicines

Y4 Wastes from the production, formulation and use of biocides and phyto-pharmaceuticals

Y5 Wastes from the manufacture, formulation and use of wood preserving chemicals

Y6 Wastes from the production, formulation and use of organic solvents

Y7 Wastes from heat treatment and tempering operations containing cyanides

Y8 Waste mineral oils unfit for their originally intended use

Y9 Waste oils/water, hydrocarbons/water mixtures, emulsions

Y10 Waste substances and articles containing or contaminated with polychlorinated biphenyls (PCBs) and/or polychlorinated terphenyls (PCTs) and/or polybrominated biphenyls (PBBs)

Y11 Waste tarry residues arising from refining, distillation and any pyrolytic treatment

Y12 Wastes from production, formulation and use of inks, dyes, pigments, paints, lacquers, varnish

Y13 Wastes from production, formulation and use latex, plasticizers, glues/adhesives

Y14 Waste chemical substances arising from research and development or teaching activities which are not identified and/or are new and whose effects on man and/or the environment are not known

Y15 Wastes of an explosive nature not subject to other legislation

Y16 Wastes from production, formulation and use of photographic chemicals and processing materials

Y17 Wastes resulting from surface treatment of metals and plastics residues arising from industrial waste disposal operations

Y18 Wastes collected from households, including sewage and sewage sludges

Y47 Residues arising from the incineration of household wastes
Wastes having as constituents:

Y19 Metal carbonyls

Y20 Beryllium; beryllium compounds

Y21 Hexavalent chromium compounds

Y22 Copper compounds

Y23 Zinc compounds

Y24 Arsenic; arsenic compounds

Y25 Selenium; selenium compounds

Y26 Cadmium; cadmium compounds

Y27 Antimony; antimony compounds

Y28 Tellurium; tellurium compounds

Y29 Mercury; mercury compounds

Y30 Thallium; thallium compounds

Y31 Lead; lead compounds

Y32 Inorganic fluorine compounds excluding calcium fluoride

Y33 Inorganic cyanides

Y34 Acidic solutions or acids in solid form

Y35 Basic solutions or bases in solid form

Y36 Asbestos (dust and fibres)

Y 37 Organic phosphorous compounds

Y38 Organic cyanides

Y39 Phenols; phenol compounds including chlorophenols

Y40 Ethers

Y41 Halogenated organic solvents

Y42 Organic solvents excluding halogenated solvents

Y43 Any congener of polychlorinated dibenzo-furan

Y44 Any congener of polychlorinated dibenzo-p-dioxin

Y45 Organohalogen compounds other than substances referred to in this Annex (e.g., Y39, Y41, Y42, Y43, Y44).

Table 4.2: Characteristics of Hazardous Wastes

Class*	Characteristics
1	1 H1 Explosive An explosive substance or waste is a solid or liquid substance or waste (or mixture of substances or wastes) which is in itself capable by chemical reaction or producing gas at such a temperature and pressure and at such a speed as to cause damage to the surroundings.
3	H3 Flammable liquids The word "flammable" has the same meaning as "inflammable". Flammable liquids are liquids, or mixtures of liquids, or liquids containing solids in solution or suspension (for example paints, varnishes, lacquers, etc., but not including substances or wastes otherwise classified on account of their dangerous characteristics) which give off a flammable vapour at temperatures of not more than 60.5 degrees C, closed-cup test, or not more than 65.6 degrees C, open-cup test. (Since the results of open-cup tests and of closed-cup tests are not strictly comparable and even individual results by the same test are often variable, regulations varying from the above figures to make allowance for such difference would be within the spirit of this definition).
4.1	H4.1 Flammable solids, or waste solids, other than those classed as explosives, which under conditions encountered in transport are readily combustible, or may cause or contribute to fire through friction.
4.2	H4.2 Substances or wastes liable to spontaneous combustion Substances or wastes which are liable to spontaneous heating under normal conditions encountered in transport, or to heating up on contact with air, and being then liable to catch fire.
4.3	H4.3 Substances or wastes which, in contact with water emit flammable gases Substances or wastes which, by interaction with water, are liable to become spontaneously flammable or to give off flammable gases in dangerous quantities.
5.1	H5.1 Oxidizing Substances or wastes which, while in themselves not necessarily combustible, may, generally by yielding oxygen, cause or contribute to the combustion of other materials.
5.2	H5.2 Organic peroxides Organic substances or wastes which contain the bivalent-O-O-structure are thermally unstable substances which may undergo exothermic self accelerating decomposition.
6.1	H6.1 Poisonous (Acute) Substances or wastes liable either to cause death or serious injury or to harm human health if swallowed or inhaled or by skin contact.
6.2	6H6.2 Infectious substances Substances or wastes containing viable micro organisms or their toxins which are known or suspected to cause disease in animals or humans.
8	H8 Corrosives Substances or wastes which, by chemical action, will cause severe damage when in contact with living tissue, or in the case of leakage, will materially damage, or even destroy, other goods or the means of transport; they may also cause other hazards.
9	H10 Liberation of toxic gases in contact with air or water Substances or wastes which, by interaction with air or water, are liable to give off toxic gases in dangerous quantities. H11 Toxic (Delayed or chronic) Substances or wastes which, if they are inhaled or ingested or if they penetrate the skin, may involve delayed or chronic effects, including carcinogenicity. H12 Ecotoxic Substances or wastes which if released present or may present immediate or delayed adverse impacts to the environment by means of bioaccumulation and/or toxic effects upon biotic systems. H13 Capable, by any means, after disposal, of yielding another material, e.g., leachate, which possesses any of the characteristics listed above.

*   As adopted from hazardous classification system included in the United Nations Recommendations on the transport of Dangerous Goods (ST/SG/AC.10/1/Rev.5, United Nations, New York 1988).

Source: OAU Secretariat 1991.

From the provision of Article 3, on becoming party to the Bamako Convention, member states are obliged to inform the secretariat of other wastes they consider hazardous. This is supposed to be done within six months. Inasmuch as the Bamako Convention stands out as a distinctive and beautiful piece of legislation, its test has been in its implementation. How many countries party to it have managed to resist the importation of hazardous wastes? This question threatened to split the African Union (then the OAU), as some countries have also ratified the Basel convention. The Waste Management Act of 1998 for Botswana (Government of Botswana 1998) makes provisions for the application of the Basel convention. Section 45 (1) outlines that the Basel Convention, including 'any amendments, appendices and resolutions thereto, shall apply in regulating the trans-boundary movement of waste'. Regulations may be promulgated in order to (Section 45 a and b): make such provision as appears necessary or expedient for the carrying out of and giving effect to the Basel Convention; and impose fees and provide for the recovery of any expenditure incurred in giving effect to the Basel Convention.

South Africa has also acceded to the Basel Convention (DEAT 2000). As one of the emerging economies in Africa, particularly in southern Africa, the fact that South Africa is party to the Basel Convention places it in the limelight, as a number of other countries in the SADC region have ratified the rival Bamako Convention. Being in the forefront of promoting NEPAD, the fact that South Africa allows the importation of hazardous wastes into its territory negatively impacts on its leading role in promoting sustainable waste management in the continent.

The ills of imported hazardous substances are probably best illustrated by events that took place at Thor Chemicals, a former British company based in South Africa's Kwazulu-Natal Province (Nyandu 2002). Thor Chemicals imported thousands of mercury waste barrels, which it failed to recycle, from the US and other countries. According to Butler (1997), Thor Chemicals was established in South Africa in 1963. Investigations into the company's activities revealed that workers had been poisoned in the process. In 1992, two workers (Peter Cele and Engelbert Ngcobo) lapsed into a coma. A third worker, Albert Dlamini was admitted into hospital after going berserk at work. The following year, Nelson Mandela raised environmental awareness regarding hazardous waste when he visited Engelbert Ngcobo in hospital in November. It was on this occasion that it emerged that Thor Chemicals was polluting the air by operating an unlicensed incinerator. Representatives of twenty affected families later sued Thor Chemicals with civil proceedings taking place in London. This resulted in the parent group (Thor Holdings) agreeing to a settlement of R9,000,000 in April 1997. Loss of memory and sexual desire, headache and, arching joints are some of the many problems that another former Thor Chemicals employee, Johannes Nxumalo reported (Nyandu 2002: 289). Although the worker was aware of the dangers associated with working for the company, poverty forced him to risk his life. Nxumalo maintains that him and his colleagues had to jog to and from work as a means of 'getting rid of the poison' in their bodies. Despite such breathtaking accounts, African governments are yet to decisively act on banning the importation of hazardous trans-boundary wastes. At a regional meeting in Nigeria in May 2004 (UNEP 2005), African governments adopted a position on the Strategic Approach to International Chemicals Management (SAICM). SAICM is seen as a tool for good environmental stewardship in the continent. Through SAICM, African governments agreed to: manage chemicals at all stages of their life cycle: the 'cradle-to-grave' principle; target most toxic and hazardous chemicals as a priority; and incorporate the principles of substitution, prevention, the polluter pays, the right to know and greening the industry; and integrate the precautionary, liability and accountability approaches. A remaining challenge is the question of harmonising the Bamako and Basel Conventions.

Conclusion

In this chapter, we discuss how the African environmental management agenda is being set. Of note is that the continent's environmental agenda is strongly influenced by donors, some of which have genuine intentions of assisting Africa to tackle its environmental ills. It also emerged that the key donors and aid agency shaping our environmental agenda are CIDA and DANIDA, and recently the African Union through the NEPAD environment action plans, as well as the African Development Bank Group's Policy on Environment. The chapter also notes that highly sensitive discourses still remain, notably how best the continent could speak in one voice in order to address some key environmental problems, such as the importation of hazardous waste. African Union members have ratified the rival Basel and Bamako Conventions. Overall, it can be concluded that Africa has embraced the concept of sustainable development as the development paradigm for the century, and that sustainability and sustainable development embrace politics, environment, economics, culture and society.

Revision Questions
1. What environmental management agenda are raised by CIDA's policy for environmental sustainability?
2. Which environmental management agenda is raised by DANIDA's environmental strategy?
3. How is the agenda different, if at all, from that portrayed by the African Development Bank Group's policy on the environment?
4. What sub-regional environmental stewardship initiatives exist in your region?

Critical Thinking Questions
1. Drawing lessons from the Bamako and Basel Conventions, on which other sensitive environmental issues have the governments of other African Union countries failed to speak out with one voice, and how best can such environmental problems and the differences between governments be addressed?
2. How best can African governments fund the environmental stewardship agenda?
3. Is NEPAD's environmental agenda adequate for the African Union?

References

African Development Bank, 2004, *African Development Bank Group's Policy on the Environment*, Tunis: African Development Bank.

Butler, M., 1997, 'Lessons from Thor Chemicals: The links between health, safety and environmental protection', in L. Bethlehem and M. Goldblatt, eds., 1997, *The Bottom Line: Industry and the Environment in South Africa*, Cape Town: University of Cape Town Press, pp. 194–213.

CIDA, 1992, *CIDA's Policy for Environmental Sustainability*, Quebec: CIDA.

Darkoh, M.B.K. and A. Rwomire, eds., 2003, *Human Impact on Environment and Sustainable Development in Africa*, UK: Ashgate, Aldershot.

Darkoh, M.B.K., 1993, 'Land Degradation and Soil Conservation in Eastern and Southern Africa: A Research Agenda', *Desertification Control Bulletin*, No. 22, pp. 60–8.

DANIDA, 2004, *Strategy for Denmark's Environmental Assistance to Developing Countries 2004–2008*, Copenhagen: DANIDA.

DEAT, 2000, *White Paper on Integrated Pollution and Waste Management*, Pretoria: Government Printer.

Government of Botswana, 1998, *Waste management Act 1998*, Gaborone: Government Printer.

NEPAD, 2002, *A Summary of NEPAD Action Plans*, Johannesburg: NEPAD.

Nyandu, M., 2002, 'Crippled for life by mercury exposure', in A.D. McDonald, ed., *Environmental Justice in South Africa*, Cape Town: University of Cape Town Press, pp. 289–91.

OAU Secretariat, 1991, *Bamako Convention on the Ban of the Import into Africa and the Control of Transboundary Movement and Management of Hazardous Wastes within Africa*, Addis Ababa: OAU Secretariat.

SADC, 1996, *SADC Policy and Strategy for Environment and Sustainable Development: Toward Equity-led Growth and Sustainable Development*, Gaborone: Southern African Development Community.

SADC, 2003, *Draft SADC Regional Indicative Strategic Development Plan (RISDP)*, Gaborone: Southern African Development Community.

SADC-REEP, 2007, *Environmental Issues and Crises*, Howick: SADC-REEP.

UNEP, 2003, *Global Environment Outlook 3: Past, Present and Future Perspectives*, London: Earthscan.

UNEP, 2005, *GEO Yearbook 2004/5*, Nairobi: United Nations Environment Programme.

UNEP, 2006, *Africa Environment Outlook 2: Our Environment, Our Wealth*, Nairobi: United Nations Environment Programme.

World Bank, 1998, *The World Bank Annual Report 1998*, Washington DC: World Bank.

World Bank, 2001, *Making Sustainable Commitments: An Environment Strategy for the World Bank*, Washington DC: World Bank.

World Bank, 2006, *The World Bank Annual Report 2006*, Washington DC: World Bank.

Chapter 5

Uncertainties and Environmental
Threats in Africa

Introduction

Africa is the second largest of the seven continents in the world, with a land area of 30,000,000 km², about one fifth of the world's total area (ECA 2001). The ECA estimates that about half of the cultivable land is arid or semi-arid, comprised mostly of desert, with the least content of organic matter. Yet, the continent remains one of the richest in terms of natural resources.

Additionally, Africa has the highest and most rapidly increasing population, with a correspondingly high population density. Factors that account for the growth and distribution of the population in Africa are ecological and historical (ECA 2001). For example, areas with ecological factors conducive to agriculture have dense populations, while those that are dry and harsh have sparse populations. The slave trade, devastating bloody conflicts and international migrations have significantly reduced the population of Africa; while colonisation and economic factors have modified the distribution of populations on the continent. The majority of Africa's people are poor, and depend on natural resources for their livelihoods.

Coupled with an ever-increasing dependence and pressure on these resources (Darkoh 1998) it is indisputable that Africa is also a region of high uncertainties and environmental problems. This chapter is dedicated to discussing local and regional environmental problems. The terms 'local' and 'regional' describe impacts, which are more or less localised, or confined within the geographical areas where they originate, as well as the impacts that go beyond national boundaries, but are restricted to the continent's boundaries. Such environmental problems include deforestation, land degradation and soil impoverishment, droughts, desiccation and desertification, wildlife de-population and extinction, and water pollution and shortages.

Deforestation

One of the most important resources, of which Africa is proudest of, is the forest. It is estimated that African forests originally covered about 3,620,000 km², a surface area about the size of India, Nepal, Bhutan and Bangladesh combined (Martin 1991). Globally, forests cover only 6–7 per cent of the earth's landmass, but provide a habitat for about 50 per cent of all known species (World Bank 1992). In line with this general assessment, Peters' (1998) work has confirmed that forests containing a large number of different species usually contain an equally diverse assortment of useful plant species, meaning that there is usually correlation between the richness of species and resources. However, he draws attention to an unusual correlation that when there is a high species diversity of plants, individuals of a given species usually occur at very low densities. What this means, for example, is that the higher the diversity of tree species in a given forest area, the lower the rate individuals of each representative species in the area, and vice versa. This is due to constraints of space: there is a limit to the total number of trees that can be packed into, for example, a hectare of forest.

Another unfortunate characteristic, noted by Peters, of the many tropical tree species is that they have difficultly recruiting new seedlings into their population, despite the abundance of pollinators, fruit and seed dispersers and fruit sets. This is so because of a number of constraints. For example, a seed must avoid being eaten, it must encounter the appropriate light, soil moisture and nutrient conditions to germinate, and it must be able to germinate and grow faster than the seeds of other species that are competing in a given micro-site. All this illustrates why it takes millions of years for the forest to be established.

The forests are important in a number of different ways. For example, they serve ecological functions such as protecting and enriching soils, providing natural regulation of the hydrological cycle, affecting local and regional climates, influencing watershed flows of surface and groundwater through water catchment area protection, and helping stabilise the global climate by sequestering carbon as they grow. In addition to these ecological functions, forests support livelihoods and confer dignity on resident rural communities. Many rural communities depend entirely on them for survival.

Unfortunately, these ecological and livelihood support functions are under serious threat, due to the rapid rate of deforestation in Africa. Deforestation refers to the removal of forest stands by cutting and burning to provide land for other purposes, such as agriculture, the construction of residential or industrial building sites, and roads, and the harvesting of trees for building materials or fuel, without replanting new trees. According to the ECA (2002), Africa is consuming its forest and woodlands at a rate only slightly lower than the deforestation rates of Latin America and Oceania, and at a rate of twice the world average.

The causes of this problem on the continent include shifting cultivation, commercial logging, harvesting of fuelwood and building materials, and pastoralism. Shifting cultivation is considered to be the major cause of deforestation (Bundestag 1990). It is a form of traditional agriculture that is widely practised in the drylands of Africa. It is an efficient way of manipulating environments that would otherwise be unproductive under arable farming (Darkoh 2003). According to Schusky (cited in Darkoh 2003), the practice is carried out by between 25 and 33 per cent of African farmers. Schusky identified two kinds of shifting cultivators: those that practise it in order to supplement other, more permanent, cropping activities; and those that have been forced out of their 'traditional' homes, with little option but to try to earn a living by encroaching on forests and other environments. To this, two other types of shifting cultivators can be added: those who are motivated to acquire large tracks of land in order to gain traditional recognition and prestige; and those who employ it as a strategy for securing sufficient pieces of cultivable land for the family, due to perceived or predicted future scarcity.

Given these four different types of shifting cultivators, there is an obvious correlation between deforestation and population growth (Barnes 1990). In the past, a sparse African population operated the system for subsistence needs (Darkoh 2003). This was characterised by long fallow periods. This explains why in some parts of Africa, such as the Machakos district in Kenya, where shifting cultivation is not the tradition, increasing population pressure leads to intensive and highly sustainable forms of land use (Tiffen et al. 1994). However, the common phenomenon today, as Darkoh (2003) notes, is that increasing population pressure leads to more intensive shifting, and a reduction in the length of the fallow period, a condition that tends to make the system unsustainable as it leads to disastrous declines in soil fertility.

Shifting cultivation is generally characterised by what is known as slash-and-burn agriculture. According to Peters (1998), the most intensive and costly way to use the forest is to cut it down, burn it, and plant something else: timber plantation, agricultural crops and pasture grasses. Cutting down and burning the forest would not only eliminate most of the biodiversity and release approximately 150 tonnes of carbon per hectare in the form of carbon dioxide and other green house gases (Keller et al. 1991); it would also increase water movement, soil erosion and nutrient loss, as well as decreasing evapo-transpiration and total ecosystem productivity (Jordan 1987). Continued burning under shifting systems ensures that woody species do not re-colonise and are, eventually, eradicated (Mannion 1995). Faced with the need to produce more food in the face of declining soil fertility, farmers often have no choice but to extend their farming activities to areas that are agro-ecologically unsuited to such forms of land use.

The effects of shifting cultivation are often associated with rapid population growth, in view of the argument that its impact was minimal when populations were low (Darkoh 2003). Although Cleaver (cited in Darkoh 2003) considers this rapid population growth under shifting systems of agriculture to be one of the prime causes of land degradation, this conclusion, according to Darkoh, is no longer tenable in all cases. Darkoh's argument is that while high levels of increasing population are strongly associated with increasing levels of poverty and pressure on the natural resource base, it is now recognised that the strong correlation does not necessarily or always imply a causal relationship running in one direction only. He uses the Machakos case, mentioned above, to illustrate that there are, indeed, other relationships. For example, (Lockwood 2000) observes that rising population densities can lead to either degradational or conservation pathways.

The latter path, that of conservation, which Darkoh (2003) describes as a Boserupian conception, explains the Machachos case. It recognises that capacity is not fixed, but can be influenced by the application of labour, investment and technology. The opposing view, that of degradation is neo-Malthusian. It states that as populations build up, land runs out, giving rise to over-cultivation and degradation. A third view, held among economists, which Darkoh (2003) designates as post-Bruntland view, down-plays the roles of poverty, population increase and land degradation in transitional societies, and highlights the importance of economic growth. This view postulates that as economic growth increases, environmental degradation is enhanced, not necessarily by poverty or rapid population growth, but by what is called the Environmental Kuznet's Curve (Kuznets 1955). The hypothesis states that in the initial phase of economic development, a country experiences increasing environmental degradation, but after attaining a certain income threshold, degradation subsides as further development continues. According to this principle, while there may be negative environmental effects as a result of poverty and rapid population growth during the early stages of economic development, these will be counteracted by later improvements in environmental quality, as incomes and living standards improve. But, as Darkoh (2003) warns, it may be unduly optimistic to apply the Boserupian or the Kuznets' paths in the hopes of successfully redressing the problems of land degradation in Africa without appropriate interventions, such as soil conservation measures and land-use policies. His argument is that the social organisation of African societies differs from those in other regions, such as Europe and South East Asia, where societies have adapted successfully to high population densities.

It has been noted that even the application of what Darkoh describes 'as appropriate intervention' might not guarantee any significant positive results. For example, in the 1990s, agro-forestry and land-use planning initiatives were widely introduced into a huge number of rural communities in Cameroon, by conservation and development projects. The intention was to reduce shifting cultivation and encourage proper land management through crop rotation. But what is the place of crop rotation in the traditional cultural landscape of the populations? Where there was fear of an outright resistance to the new practice, projects started by encouraging what was described as selective burning. This was the practice of gathering the destroyed tree biomass into a few heaps before burning when materials are dry, so as to considerably reduce the surface area affected by fire. But farmers found this not only time-consuming, but also difficult to apply in a majority of circumstances. However, they adopted the agro-forestry option, as they could appreciate its benefits. Of course, agro-forestry has been an integral part of their traditional culture, although the 'introduced' version of planting certain (mostly leguminous) species of trees on their farms could easily be seen as slightly different from their practice of discriminatory felling, aimed at leaving behind trees of economic value. It should also be noted that in some cases, the latter practice was adopted either because the farmers did not have the appropriate implement, or because it was not financially viable to hire someone else to cut down certain (mostly large) trees.

In Kenya, the government initiated several tree planting campaigns. However, by the late 1940s, these campaigns had become immensely unpopular because they represented a cornerstone of the colonial government's land policy (FAO 1987). Although some communities adopted the 'new' version of agro-forestry, this did not prevent them from continuing with their traditional shifting cultivation. For example, in the late 1920s, the colonial forest service introduced a village forestry scheme in Malawi to promote the protection of indigenous forests, placed under the jurisdiction of village leaders. This initiative was motivated by the traditional practice of communities setting aside some forest areas for protection. The expectation was that since this initiative was in line with, and promoted, traditional practice, it would receive community support. But, instead, it was viewed with suspicion. Why? The simple reason was that the implementers had failed to consider the cultural motivation behind the practice.

Similarly, even if it were popular, the fact that it was placed under the jurisdiction of village leaders would have posed another kind of problem. For example, what could have been presented to the farmers as a disincentive against the drive to gain recognition and prestige for securing large tracts of land for future use by growing numbers of extended family members? All in all, as the FAO notes, some of the problems with tree planting interventions in the past were due to the fact that local people resented foreign initiatives, especially as these brought back memories of some of the negative results of colonialism. Local people distrusted government motives, for fear of further land alienation.

Logging

Second to shifting cultivation as a cause of deforestation in Africa is logging. This activity is largely promoted by high foreign demands, due to the extremely high rates of consumption by the industrialised nations (Struhsaker 1998; Durning 1992). Added to this is the fact that tropical countries often struggle under massive debts that drain their viability and encourage them to liquidate their forest capital more quickly, in order to raise foreign exchange. Similarly, logging offers an easy means of providing access roads in rural areas, and easily wins rural community and national support. However, it has been noted that commercial tree felling has produced a notable impact on the forest ecosystem. The physical evidence of this disturbance is immediately apparent. It persists for many years in the form of logging roads, skid trails and scattered stumps (Peters 1998).

To reduce the rising problem of deforestation due to logging, many exporting countries now insist on timber companies processing wood before it is exported. This not only provides another source of employment for citizens, but also tends to 'delay' the destruction of the forests, as time is divided between forest logging and wood processing. Furthermore, these companies are often given twenty to thirty years to carry out logging operations in their concessions, usually divided into plots.

The rule is that each plot is to be logged within a year, in a defined order. No company is allowed to move to the next plot within the same year, whether or not there is enough wood in that year's plot (Efansa Sunny 2003, personal communication). Also, by specifying the DBH (diameter at breast height) of every timber species that a timber company is authorised to harvest, the aim is to set some limits on logging, with a view to avoiding more serious forest damage, and encouraging forest management. Other practical measures have also been proposed to reduce the impact of logging operations. For example, companies are usually heavily taxed (Enviro-Protect 1997); and a percentage of the tax money is set aside for reforestation of the exploited area.

High taxes could control the number of companies able to operate in a country. But, as Enviro-Protect (1997) observes, the tax money is often either not well managed, or the regeneration effort is instead concentrated on the planting of exotic species that present a threat to the genetic diversity of the area. A combined effort of establishing tree plantations, improving forest management practices, and getting more value out of existing wood resources is required to curb the prevailing unsustainable logging of tropical forests, and to help developing countries meet growing demands for wood products (Postel and Heise 1988). The practice of selective logging has been espoused as a sustainable logging method. This involves selective cutting and removal of only the most desirable timber species. But, as Peters (1998) notes, this practice is also known to produce a number of ecological repercussions, although it is generally less damaging than total forest conversion. In tropical forests, crowns of many large canopy trees are lashed to those of neighbouring trees by a profusion of vines, lianas and clumbers. When selected timber trees are felled, other canopy tree species are pulled down and the whole woody mass crash through the lower canopy to cause even more damage. In the final analysis, harvesting even a small number of stems can destroy up to 55 per cent of the residual stand, and seriously damage an additional 3–6 per cent of the standing trees (Johns cited in Peters 1998). There is also the strategy of 'debt-for-nature' swaps proposed by conservationists. Here a conservation organisation purchases a given amount of debt owed by a timber exporting country at a discount. Taking advantage of the local currency exchange, the money is used to fund conservation efforts.

Harvesting Fuelwood

There is a flood of literature that has largely blamed deforestation in Africa on the harvesting of fuelwood and building materials by rural communities. Although this allegation cannot be totally dismissed, it did not, as it does not, reflect the real picture of the situation in comparative terms. Darkoh (1998) offers a comprehensive comparative analysis of the *status quo*. Based on 1989 estimates, the Office of the UN High Commission for Refugees found that roughly 11,000,000 trees were cut for shelter needed during the initial period of refugee influxes in Africa (Cardy 1994). This represents deforestation of over 12,000 hectares for that year. In addition, approximately 4,000,000 tonnes of fuelwood were consumed by refugees in Africa in the same year. Taking a more general view, there were widespread reports of the destruction of woodlands as a result of clearing for fuelwood. However, Darkoh (1998) observes that while the incidence of deforestation resulting from fuelwood requirements can have some serious effects, because fuelwood and charcoal are critical resources for the poor, recent research has revealed that these effects tend generally to be localised in the dryland areas, especially around settlement nodes. It has also been generally observed that most local people collect only dead wood of selected species to satisfy their fuelwood requirements. UNSO (1992) notes that in a purely rural setting, with dispersed settlements, fuelwood has seldom been a big problem. In line with this, Darkoh notes that, with the exception of some highly localised or highly populated rural areas, such as the Ethiopian Highlands and parts of Central Tanzania, little evidence exists to suggest that rural household fuelwood consumption is responsible for large-scale deforestation in Africa. Accordingly, mounting evidence, notably from Zambia, Kenya and Sudan, points out that it is rather urban demand, usually for charcoal, that leads to the extensive cutting down of forests. It should also be noted that high charcoal production is often carried out in areas where logging operations are extensive, as locals collect waste-wood free of charge, or at low cost. Urban charcoal demands are also evidently high in Cameroon and Nigeria where some 'modern homes' additionally construct what is known locally as firewood kitchens. This is not only a strategy to reduce the cost of consuming the rather expensive commercially produced gas, but it is also necessitated by the belief that certain traditional dishes would not taste as good if they were not cooked on biomass stoves.

Several methods have been proposed to combat deforestation due to the fuelwood crisis and increasing demand. These include publicly-supported tree growing programmes, community management of natural forests, and the use of improved biomass stoves. The first two strategies constitute what FAO (2003) identifies as the two models of wood production. Ethiopia provides a historical example of a publicly initiated tree growing programme to promote the production of fuelwood (FAO 1987). In the late 1890s, the Ethiopian Emperor Menelik introduced legislation to exempt land planted with funds from taxation. Furthermore, he arranged for the distribution of eucalyptus seedlings at nominal prices. All this was a response to an extreme scarcity of wood around the then new capital of Addis Ababa. Although the programme was slow to start, by the 1920s the streets and paths of Addis Ababa were already looking like a vast continuous forest. Apart from its use as fuelwood, people found other uses for eucalyptus, such as house construction, implements and furniture. So the activity spread to other parts of Africa, mostly more sparsely forested areas. However, today the eucalyptus tree is gradually losing popularity, due to its observed water-losing rather than water-conserving capacity; however, it continues to be maintained wherever it was planted. The spread of neem trees in West Africa and the Sahel, introduced into Senegal in 1944 and into Mali in 1953, is another striking example of the successful introduction of a new tree species in the more recent past. The eucalyptus and neem tree species are highly valued for both their rapid growth, and multi-purpose uses.

Similar to tree planting, there is increasing emphasis in the continent on what is popularly known as community forestry. Community forestry could be defined as any situation which immediately involves local people in a forestry activity, in areas which are short of wood and other forest products, through the growing of trees at farm level to provide cash crops and other income generating products, large-scale industrial forestry, and any other forms of forestry which contribute to community development. On-farm community forestry has taken the form of agro-forestry activities that have been widely promoted, technically and financially, by the conservation and development projects that have proliferated on the continent. Tree nurseries have been established and seedlings distributed to community members, free-of-charge or at very minimal costs. On the other hand, management of natural forest areas, in what are now popularly known as community forests, has been done with the technical assistance of governments, though often with the financial support of conservation and development projects. According to FAO, they are generally designed to use land under direct community ownership, or state land, which has been specially designated for community control. Unlike farm-based community forestry, control of which is by the individual farmers, in this situation management is either by the community as a whole, or through a community group.

Perhaps the best known community forest initiative, albeit outside of Africa, has been the Chipko Movement in the Himalayan region of Northern India. It started when a large group of people with common interests in tree resources saw that by organising themselves into a group, they could more effectively influence the political and economic forces that were improperly managing their physical environment. This movement first came to public attention in 1973, when members demonstrated against the commercial felling of trees in forests where they lived by hugging the trees to stop fellers from cutting them down.

Despite the acclaimed success of community forestry in some parts of the continent, there are many instances of less encouraging results from such programmes in other parts. For example, we have seen the obvious failures of tree-planting campaigns in Kenya and Malawi (FAO 1987). Specifically, many agro-forestry programmes have failed for the same or other reasons. One other reason is the progressive reduction in tangible benefits to farmers, especially when there is an increasing surplus of supplies, and prices are low. Another important reason for failure is related to the labour-intensive nature of the activity, vis-à-vis normal farming operations. Also, since such projects are often imposed to meet mostly long-term objectives, rather than desirable short-term community objectives, participants begin to feel sooner or later that it is merely a wasted effort on their part to continue. One popular project management recommendation today is to avoid undue introduction of new practices, but rather to build on or, at worst, fine-tune indigenous systems in order to make them culturally acceptable, socio-economically realistic and viable, and, therefore, sustainable.

According to FAO (2003), local management of natural forests is also hampered by weak and slow-changing institutions, rent capture by local elites, inconsistent laws and regulations, and a cumbersome bureaucracy. In some cases the government has refused to give full ownership and control of community forests to communities. It may seem logical to conclude that the concept of participatory forest management more appropriately describes what are generally designated as community forests by most African governments. Participatory forest management is all-encompassing. It includes different types of forestry activities that involve local communities exercising different levels of decision-making authority. Concepts such as community forestry, community-based forest management, social forestry, joint forest management, collaborative forest management, common property forest management, and participatory forestry all refer to approaches that involve local communities at various levels. The use of such a broad definition is important because it embraces the experiences from countries that differ in their approaches.

Improved biomass stoves are fast becoming popular. They are seen as an effective means of reducing fuelwood consumption due to their heat retention capacity. This type of appropriate technology has the obvious advantage, over and above using very little wood at once, of encouraging the use of what would have otherwise been wasted. Although many samples of high quality stoves are available on the market, there is the need for further research to increase their quality in terms of heat retention and smoke emission (Eyabi 2005, personal communication).

Pastoralism

Pastoralism is another cause of deforestation in Africa. It is a form of agriculture that is carried out in lands that cannot be readily cultivated to produce crops, or where it makes economic sense to generate animal products. It may involve cutting down forest areas and replacing them with pasture grasses. Nomadic pastoralism occurs in arid and semi-arid environments, which necessitates the migration of animals and herders on a regular basis from one area to another. Like shifting cultivation, nomadic pastoralism, when well managed, is ecologically balanced through the varying requirements of livestock components. This makes the agricultural system stable in a fragile environment. Grazing, however, becomes a threat to biodiversity because it not only displaces wildlife but also disrupts the composition of plant species. When the population increases, there tends to be a permanent replacement of naturally occurring biota, with lasting changes in the output, input and components of the system. This system is associated with a major reduction in species diversity, as just a few domesticated animals replace the wide range of wild species (Mannion 1995). Farmers often remove trees from grazing lands with the intention of improving grass growth by reducing competition for water and soil nutrients. But this has an adverse effect on biodiversity because of tree loss.

Land Degradation and Soil Impoverishment

Although there is a distinction between the two concepts of land degradation and soil degradation (or soil impoverishment), they both describe a decline in land or soil quality due to human activities. Darkoh (2003) defines degradation as diminution or destruction of the biological potential by one or a combination of processes acting on the land. The concept of land degradation is broader than soil degradation, because it deals with the whole ecosystem, of which the soil is one component. The 1994 UN International Convention to Combat Desertification described land degradation as the reduction or loss of the biological or economic productivity and complexity of the land (INCD 1994).

When land degradation is a result of human activities, it usually arises from a mis-match between land quality and land use (Beinroth et al. cited in Darkoh 2003), with attendant implications for land productivity. Constant exposure of a piece of land, through, for instance, forest clearance, bush burning and over-grazing, destroys the soil structure and texture, as well as the useful organisms responsible for its formation. Land and soil degradation are also due to over-cultivation, which results in the depletion of plant nutrients, as well as inappropriate or over-application of fertilizers, poor irrigation and other inappropriate land-use practices.

The processes of land degradation include water erosion, wind erosion and sedimentation; long-term destruction of vegetation, diminution of many plants and animal populations; or decreases of crop yields, salinization or sodication of soils (Darkoh (2003). Lal (1994) identifies, in broader terms, three key mechanisms that initiate land degradation: physical, chemical and biological processes. Physical processes include a decline in soil structure leading to crusting, compaction and erosion. Chemical processes include acidification, leaching, salinization, decrease in nutrient retention capacity and fertility depletion. Biological processes include reduction in biomass carbon and decline in land biodiversity. Physical soil degradation results in the deterioration of the structure of the soil, making it more compact and harder to use, due to impermeability to rain and poor drainage. The soil also develops hardpans and surface crusting (ECA 1999). Soil structure is the important property that affects all three degradational processes.

Soil degradation is more pronounced where the soil is exposed to agents of denudation and erosion. Erosion has been defined as the washing away of the topsoil, together with its valuable plant nutrients. Wind and water erosion types are extensive in many parts of Africa, with an estimated 25 per cent of the non-desert landmass prone to water erosion and 22 per cent prone to wind erosion (WMO cited in ECA 2002). Using South Africa to illustrate the gravity of erosion, water erosion alone affects 6,100,000 hectares of cultivated soil, 15 per cent of which is seriously affected, 37 per cent, moderately, and the rest slightly. Wind erosion is more severe as it claims an estimated 10,900,000 hectares of cultivated soil, 7 per cent of which is seriously affected, 29 per cent, moderately, and 64 per cent, slightly (Landcare Supplement cited in ECA 2002).

As in the case of deforestation, high population density is readily considered the root cause of soil degradation in Africa. For example, Darkok (2003) argues that increasing population pressure leads to more intensive shifting, and to a reduction in the length of the fallow period, to such a degree that the soil is no longer allowed to replenish its nutrient store; hence a decline in soil fertility. However, although increasing population density cannot be wholly dismissed, it is not the root cause of land degradation in Africa. Rather, the extent of degradation is determined by what a population does to the land (ECA 2002). For example, poor land-use practices, such as slash-and-burn agriculture, over-cropping and other farming methods, and the inappropriate use of fertilisers, have deleterious effects on soil fertility. According to the ECA, land degradation is closely linked to land tenure systems. This is evident in the fact that if a people do not own the title to the land, they have no incentive to invest in long-term improvements. Similarly, due to the question of ownership, traditional methods of managing grazing have become less effective. The consequence is that free-range grazing has led to overgrazing, especially in arid and semi-arid areas, resulting in deteriorated land cover. It could, therefore, be concluded that land degradation is a biophysical process driven by socio-economic and political causes (ECA 1999).

The picture of soil degradation in Africa is frightening. In sub-Saharan Africa, 72 per cent of arable land, and 31 per cent of pasture land, are degraded. The rate of soil loss in South Africa is estimated at 400,000,000 tonnes annually (Griffen cited in ECA 2001). An estimated 30 per cent of smallholder farmland in Zimbabwe is now totally degraded. The situation is worse in the densely populated areas of Malawi, such as the Lilongwe plains. It is estimated that 14 per cent of degraded soil results from vegetation removal, 13 per cent from over-exploitation, 49.5 per cent from overgrazing, and 24 per cent from agricultural activities (WRI cited in ECA 2002).

Land degradation was a major global issue in the twentieth century. It will remain high on the international agenda for many more decades to come. This is due to its impact on world food security and the quality of the environment (ECA 2002). When soils are subjected to severe and extreme degradation, their original biotic functions are damaged. Reclamation is, at worst, impossible and, at best, difficult or costly (FAO cited in Darko 1998). The resulting degradation of productive lands has led to declining production and intensified food insecurity (Darkoh 1998).

Several measures have been taken to address land degradation in Africa, ranging from modern to traditional methods and techniques. Modern methods, promoted by governments and NGOs, include improving farming methods and systems, with specific emphasis on agro-forestry for soil enrichment, or conservation and erosion control. Practical steps have been taken to introduce modern cultural practices, such as terracing and 'horizontal ridging' on slopes.

In the case of traditional methods, Leach and Mearns (cited in ECA 2002) have shown that in West Africa, counter to assumptions about environmental destruction, indigenous rural land users have maintained and increased environmental productivity. This has been achieved through maintaining a balance between the amount and constituents of forest and retained grassland cover, using traditional land management techniques, which respond to changing socio-economic environments.

Another striking example of traditional methods is the Machakos experience in Kenya. Tiffen et al. (1994) investigated the processes leading to the recovery of the landscape of Machakos District. It was observed that for a period of sixty years, landscape degradation had been followed by the improvement of the same landscape, despite a six-fold rise in human population. The recovery of the landscape was accompanied by and linked to marked improvements in human welfare. The improvements in physical and human environments had been achieved through a complex blend of externally generated technical innovations and economic changes. But their successful adoption has been due to indigenous initiatives. Toulmin (1995 cited in ECA 2002) also notes that the Machakos miracle was created by ordinary people adapting to livelihood opportunities, thus underscoring the importance of applying indigenous knowledge when addressing the environment.

Droughts, Desiccation and Desertification

Africa's dryland environment poses formidable problems for sustainable development. Amongst these are unpredictable and severe drought, and desiccation or aridification due to persistent drought and dryland degradation and desertification. Darkoh (1998) offers clear definitions of desertification, drought and desiccation, terms, which there is a tendency to confuse. He defines 'drought' as a naturally occurring short-term phenomenon, when precipitation is significantly below normal recorded levels. It is a dry period from which an ecosystem recovers rapidly after the rains return. Such temporary deficits in rainfall can generally be accommodated by existing ecological, technical and social strategies. 'Desiccation' or 'aridification' is defined as a longer-term deficit in rainfall, which seriously disrupts ecological and social patterns. It requires national and global responses. Recovery after desiccation is much slower, because trees may have died and vegetation may take many years to recover.

Several definitions have been advanced for desertification. The most current and widely accepted is that used by UNCED (1992). According to UNCED, desertification is 'land degradation in arid, semi-arid and dry sub-humid areas resulting from various factors including climatic variations and human activities'. An important limitation is that desertification is restricted to the drylands. The definition identifies two key factors responsible for desertification: 'climatic variation' and 'human activities'. 'Climatic variation', or 'climate change', refers to short-term climate variability, and longer-term climatic trends or shifts caused by natural mechanisms, or by human activity (Kelly and Hulme 1993).

Meanwhile It is important to note that the human activities responsible for desertification are not different from those that cause land degradation. They include over-cultivation, over-grazing, poor irrigation and other land-use practices.

Desertification is effectively land degradation restricted to dryland areas. Similarly, there is an obvious correlation between drought and desiccation, and desertification. However, as Darkoh warns, it does not necessarily follow that drought and desiccation will give rise to desertification. His argument is that this is only likely to be the case when human misuse or mismanagement of the land weakens the natural environment (Darkoh 1998). Whether desertification is likely to occur or not depends on the resource management in the dryland areas. Therefore, climate variations alone, characterised by droughts and desiccation, are unlikely to result in desertification where there is proper resource management, as this should work against the natural factors.

Desertification has been mistakenly understood as shifting sand dunes, as expressed in the widely used term 'desert encroachment'. The common feature of a desert is hence understood as a large expanse of barren rock or sand, with very sparse vegetation. This misconception originates in concerns about shifting sand dunes and sand drifts, where wind erosion is particularly serious in cultivated areas in which dry farming is practised (Darkok 2003). UNCED (1992) has dispelled this misconception by stating categorically that the process of desertification is not due to shifting sand dunes, but rather 'patches of increasingly unproductive land breaking out and spreading over hundreds of square kilometers'. Desertification may take centuries. However it should be noted that once a desert has formed, it is very difficult to restore it as arable land.

According to WWF (1988), deserts are spreading over 6,000,000 hectares of land every year. This is a very serious situation, given the estimate that 73 per cent of the total drylands used for agriculture are degraded (Table I), and a third of Africa is affected by desertification. Three distinct areas of the continent are at most risk: Mediterranean Africa, the Sudano-Sahelian region and the Kalahari-Namib region in southern Africa (Darkoh 1998).

Drought, desiccation and desertification have dire consequences for the continent. Darkoh (1998) reports that the great drought in the Sudano-Sahel region in the early 1970s claimed about 250,000 lives. This was actually less severe than the drought of 1982–5, which affected the entire sub-Saharan region. Ethiopia was the worst affected, with an estimated 1,000,000 people starving to death from the combined effects of drought and civil war. Drought reduces millions of people to destitution, driving mass migration to urban areas in search of work and relief. Additional pressure is placed on basic city services such as water and sanitation, which introduces another set of problems. The agro-economic effects of drought include lower and more variable yields, reduction in acreage of cropped lands, less high-yield food cropping, diminished rangeland productivity, changes in pastures and in the composition and size of herds, and lower prices, as herdsmen flood the market with sickly cattle, seeking to sell them before they die.

The effects of desiccation on croplands and rangelands have been much more serious than those of droughts. In the Sahel, many peasant and pastoral communities have 'simply ceased to exist after the desiccation of the last 20 years' (UNSO

1992:30). As would be expected, the effects of desertification are even more pronounced:

> Desertification reduces the productivity of land and deprives people of biological resources that are important for human sustenance. These impacts, in turn, lower incomes (and subsistence levels) of hundreds of millions of already poor, dryland peasants, herdsmen and urbanities who form part of the same economy. Prolonged periods of drought under these circumstances lead to hunger, malnutrition and starvation, high infant mortality and accelerated rural migration. Loss of biodiversity in cultivated plants and domesticated animals, and in wild foods which are so important when agriculture fails at times of drought, is a direct threat to food security (IPED 1994 cited in Darkoh 1998:13).

> …[It] translates into a spiral of declining production, increasing poverty and diminished potential productivity. The exacerbated poverty, in turn, exacerbates desertification because, as the pressure increases, people are forced to exploit their land to survive. In doing so, they further diminish its productivity and the cycle continues. The result is seen today in the Ethiopian Highlands and all across the Sudan-Sahel: starvation, death, disease and the exodus of millions of environmental refugees moving in desperate search for survival to the urban areas or to other less degraded lands elsewhere. Directly or indirectly desertification slowly erodes the genetic base for human staple food and undermines the whole production system. Entire societies and cultures are now threatened. The pastoralists of the Sahel are a case in point. For most, the loss of their livelihood means a life in relief camps or in shanty towns mushrooming around Sahelian cities and those of the countries to the south' (Darkoh (1998:15).

Ghai (1992) adds that desertification and resource-scarcity may provoke social unrest and political and armed conflict. Several governments, most notably the Haile Selassie regime of Ethiopia, have been swept from power due to the suffering and unrest associated with drought and famine. With continuing degradation and the increasing scarcity of natural resources, the struggle and competition for the remaining resources are likely to become a potent source of conflict among communities and countries in the African drylands

Wildlife De-population and Extinction

Wildlife de-population describes a considerable drop in the population of wildlife, due largely to the excessive exploitation of species, and the uncontrolled habitat destruction by humans. Exploitation of wildlife for commercial purposes (e.g. hunting and logging) renders the future of some wildlife species rather uncertain. It is important to note that many rural communities largely depend on the sale of game meat for income. Wilcox and Nzouango (2000) carried out hunting studies at the Banyang-Mbo Wildlife Sanctuary in southwest Cameroon. They recorded a total kill of 8,139 animals in eighteen months. This increased rate of hunting is due to a combination of high population growth, improved hunting techniques, and a high demand for game meat. The meat is usually smoked and sold in local markets on a weekly basis. With increasing external demand, a commercial network is developing to expedite movement of this product from the rural area to the urban area, where game meat is an important source of protein (Robinson 1996).

Wildlife–human conflict is another cause of the severe declines in wildlife numbers, due to increased poaching, as communities consider shooting to be the most effective method of addressing the problem (Douglas-Hamilton 1997; Inyang 2002). This conflict may be attributed in large measure to a combination of rapid human population growth and poor land-use management strategies, such as shifting cultivation, which impose increasing demands on the land. This demand results in ever-increasing encroachment on wildlife habitats for agricultural activities, and the development of human settlements. The situation is clearly explained by Hunter (1996) who postulates that the geometry of natural habitat fragmentation, induced by agriculture, indicates that as the wildlife range contracts in the face of human expansion, the interface of potential wildlife-human contact increases. Inyang (2002) observes that even the hunting that Banyang-Mbo in Cameroon communities often employ, as a strategy for resolving conflict (resolution through the elimination of key individuals), instead perpetuates conflict, as the targeted species are forced to disperse wildly and cause more widespread problems due to increased disturbance.

The destruction of valuable wildlife habitats to make way for plantations and human settlements is a serious threat to the wildlife. A drastic drop in the population of a species may render it endangered. A species is considered endangered when it is faced with extinction or when it is in danger of becoming extinct. There is a long history of why certain species became endangered.

One way of controlling wildlife de-population is to stop the unsustainable exploitation of species, as well as the indiscriminate destruction of their habitats, through a combination of environmental education, livelihood support activities and law enforcement. Environmental education should form the greater part of the effort, and law enforcement should be a last resort, mostly effective if and when there assured continued collaboration of the target population. Reducing the consumption of endangered species through the introduction of supplementary sources of income and protein offers sustainable redress. But this approach can be deceptive, because preferences are influenced by a combination of factors that include fashion, taste, cultural values and taboos, production cost, and price differences. This approach presupposes availability to be the only economic factor that could influence such preferences. But it should be noted that these are substitution preferences, differing from one individual to another. They are often coloured or complicated by such economic variables as scarcity, elasticity of demand, elasticity of supply, and opportunity cost.

Moreover, some cultural taboos have a far greater influence on individual choices than economic factors. For example, most people from some ethnic groups in the North West Province of Cameroon still hold the firm belief that by eating snails, they would develop scabies all over their bodies, among other fearsome conditions.

No amount of persuasion, or motivation, would entice them to contravene or violate their beliefs. Unfortunately, some project-implementers may be excited about the idea of introducing snails or some other alternative livestock, having learnt about their success elsewhere.

Alternatively, some of the alternatives introduced may satisfy all the cultural factors, as well as taste, but not the economic factors. For example, the introduced livestock may eventually be found to cost more than what it was intended to out-compete, due to the high cost of production. In this case, the income level of the community members would become the determining factor. It could therefore be concluded that for new livestock to be introduced, it must be both culturally and economically acceptable. Therefore, a series of studies must be conducted before deciding what introduction would satisfy both economic and cultural factors, i.e., what has the potential to place itself high on the scale of the communities' preferences.

Extinction

As a result of continued wildlife de-population, many valuable plant and animal species are being lost. This happens mostly before their uses have been discovered, as a result of excessive exploitation of the earth's flora and fauna and the un-checked destruction of wildlife habitats. Scientists estimate that every hundred years, one species, either a plant or an animal, becomes extinct naturally. But today an astounding number of species become extinct every year, due to a wide range of unfriendly and outright destructive human activities.

Extinction is the end-result of wildlife de-population. It may be local or global. Local extinction is the loss of a species in a particular region, while global extinction describes the complete extermination of a species from the earth. Local extinctions are a common phenomenon in fragmented habitats and islands. Habitat fragmentation could be brought about through road construction and agricultural activities. The resultant fragmented blocks share similar characteristics with islands. Local extinctions in these habitat areas can be countered by re-colonisation from larger habitat blocks, in the case of habitat fragments, and from the mainland or other islands, in the case of islands. But in the case of endemic species on a remote island or habitat fragment, re-colonisation is impossible. This type of local extinction is therefore precisely equivalent to global extinction (Begon et al. 1996), which is an irreversible phenomenon. Normally, local extinction could be addressed through breeding and reintroduction programmes. But a solution to global extinction is yet to be discovered: there simply is no solution. The loss of a species may have great ecological, economic and cultural consequences.

Ecologically, the removal of a species, especially a keystone species, which may be a top predator or an important prey, could disrupt the ecosystems that perform a wide range of functions, from nutrient recycling to global climate moderation.

Economically, some species provide direct sources of income for local communities. Indirectly, they provide a free source of materials for food, building and medicine. Culturally, some rural communities are said to have spiritual relationships with certain species, the removal of which could mean the collapse of such links.

To prevent extinction requires banning the trade, exploitation and purchase of endangered species. Another strategy is an expanded programme of environmental education aimed at enabling the target populations to identify the problem, appreciate its magnitude, analyse its causes, identify possible solutions, distinguish between the species in question and those that are not affected, and appreciate the roles they can play in addressing the situation.

Water Pollution and Shortages

This is the contamination of seas, rivers and other water bodies with refuse, human and industrial wastes, or pesticides and fertilizers (Figure 2.11). There is also a common practice by some local communities in, notably, some parts of Cameroon and Nigeria, where people use gamalin 20 or gamalin 40 (insecticides) to kill fish in rivers. This is a serious form of water pollution, resulting in the massive destruction of aquatic life. People eating the fish may easily become sick and eventually die due to the poison. Others may suffer long-term effects, such as infertility or abnormal births.

The excessive use of the now banned pesticide known as DDT (dichlorodiphenyltrichloroethane) produced devastating effects on birds, to the point where some, the dodo for example, became extinct. The pesticide was used to launch an effective campaign against the malaria-transmitting anopheles mosquitoes. But it was later discovered to be responsible for the thinning of the birds' eggshells. Because the eggs were thinner than normal, they easily broke whenever the birds sat on them to effect incubation.

Many soils have become degraded to such an extent that farmers who can afford to do so often use fertilisers. But often, as Cunningham et al. (2003) note, many farmers over-fertilise because they are unaware of the specific nutrient content of their soils, as well as the nutrient requirements of their crops. Large amounts of phosphates and nitrates contained in the fertilisers become a major source of aquatic ecosystem pollution. These nutrients may find their way into the aquatic system and cause cultural eutrophication or enrichment, a condition, which favours 'algae bloom' (the rapid multiplication of algae). When the algae decay, they choke off oxygen and sunlight required by other aquatic plants and animals (French 1990), which find themselves competing for the fast reducing dissolved oxygen. This phenomenon is termed biological oxygen demand (BOD). It has very serious consequences for aquatic life.

A similar, but worse, situation is when there is an oil spill, such as that which happened at Santa Babara in 1969. This may require a clean-up effort to ameliorate the situation. But the damage is already done, because when the oil spills, oxygen is prevented from dissolving into water since it cannot even penetrate through the oil layer. Considering the fact that most living organisms cannot survive without oxygen for a very short time, this produces a very serious and unfortunate situation that can cause untold devastation of aquatic life within just a few minutes.

Farmers using pesticides wrongly is another serious source of water pollution, as these chemicals are eventually washed into streams and rivers. The chemicals contain poisonous substances that not only destroy aquatic life, but may be stored in their organs, undergoing constant recycling through the food chain in a process known as bio-magnification. When these chemicals finally reach humans at the top of the chain, their levels are usually so highly concentrated that they constitute serious health and reproductive problems.

Addressing the water pollution problem, especially at a community level, requires the proper education of communities about health and other effects. Traditional injunctions have been used in some rural communities; these have proved to be successful. At a national level, there is need for intensified law enforcement to restrict the movements of harmful and commonly abused products. If many countries increased their enforcement policies, manufacturers might be forced to cease producing such products, due to the loss of markets. In other words, international laws should be strengthened, duly backed by the appropriate political will, to ensure the effectiveness of the international convention; the objectives of which are to fight against the trade in such products, and the discharge of toxic wastes.

The problem is so serious that water, once a freely available common access resource, has become scarce. It is difficult to access, and is often an expensive commodity in many countries (ECA 2002). Although water scarcity is defined more in terms of its physical unavailability, more accurately, water is available, but is polluted, and therefore scarce because it is not safe for drinking. Other factors that cause water scarcity are natural: such as a prolonged dry season or frequent droughts; others are anthropological activities, such as catchment destruction. The problem of water scarcity is severe on the African continent, and the consequences are many. As ECA (2002) notes, lack of access to good quality water has far-reaching impacts on the social, economic and environmental security of African communities, with the most seriously affected being the poor.

In rural Africa, about 65 per cent of the population do not have access to an adequate water supply. Water demand in the SADC region is projected to rise by at least 3 per cent annually until 2020, in line with population growth projections (ECA 2002). Consequently, it is estimated that by 2025, up to 16 per cent of Africa's population will be living in countries facing water scarcity, and 32 per cent will live in water-stressed countries (WWF 2000 cited in ECA 2002).

With the increasing demand for water, water-poor countries are looking to cross-border sources for future supplies. In most cases, such countries try their best to store large quantities of water in dams, thus altering and reducing the natural flows of rivers. The dams themselves do not provide any guarantee of water security for the host countries, as their future depends largely on land-use patterns in the neighbouring countries. For instance, soil erosion in the river catchments, due to activities of neighbouring countries, will cause siltation, thereby reducing water quality and the viability of dams. Thus, a dam is affected 'upstream' by a neighbour, which in turn, causes impacts on a 'downstream' neighbour (ECA 2002). In a number of African countries, as an alternative to dam building projects, efforts to address the problem of water scarcity, largely by local NGOs, have concentrated on the development of bore-holes and wells in rural communities, with the financial assistance of international NGOs, and foreign embassies. Although there are many success stories, such initiatives do not in themselves provide long-term water security for the beneficiary communities, because the sustainability of such projects depends not only on the water management potentials of communities, but also on a reduction in the degradation activities that have implications for water table levels.

Conclusion

This chapter has set out the different types of environmental problems in Africa. Although these problems are generally due to human activities, some are more closely associated with a lack of development, with poverty as the root cause; while others are linked with unsustainable economic development, with over-consumption as their root cause.

It may be summarised that environmental problems are generally due to two major factors: resource depletion and environmental pollution. Resource depletion, leaving aside other contributory factors, encompasses deforestation, land degradation, wildlife de-population and species extinction. Problems associated with environmental pollution include air, water and land pollution, global warming, acid deposition and ozone layer depletion.

Revision Questions

1. With at least three examples for each, distinguish between local and global environmental problems.
2. Name three environmental problems, analyse their causes and propose solutions.
3. Discuss some measures used to check logging activities in Africa and why it has remained one of the major causes of deforestation.
4. What factors contribute to wildlife depopulation in Africa?
5. Land degradation is at the root of hunger and starvation in some parts of Africa. Discuss the contributory factors and propose measures to ameliorate the situation.

Critical Thinking Questions

1. To what extent is shifting cultivation a major cause of deforestation in Africa?
2. Tropical rain forests are disappearing at rates that threaten the economic and ecological functions they provide. Discuss.
3. What is the correlation between deforestation and global warming?

4. Wildlife-human conflict is another cause of severe declines in wildlife numbers. Expatiate. Air pollution and acid deposition are problems associated more with industrialized nations. Why should an African worry about these problems?
5. When can the saying 'Global warming is global warning' be justified?
6. Nuclear technology is both a blessing and a curse to the world. Discuss.

References

Begon, M., Harper, J.L. and Townsend, C.R., 1996, *Ecology*, London: Blackwell Science Ltd.

Bundestag, D., 1990, *Protecting the Tropical Forests: A High Priority International Task, Second Report of the Enquete-Commission*, 'Preventive Measures to Protect the Earth's Atmosphere', of the 11th German Bundestag Retreat Offentlichkeitsarbeit.

Cardy F., 1994, *Environment and Forced Migration*, Nairobi: UNEP.

Cunningham, W.P., Saigo, B.W. and Cunningham, M.A., 2003, *Environmental Science: A Global Concern*, New York: McGraw-Hill.

Durning, A., 1992, *How Much is Enough? The Consumer Society and the Future of the Earth*. New York: W.W. Norton and Company.

ECA, 1999, *Study on Soil Erosion and Destruction of Land Resources: Issues and Trends in Africa*, Addis Ababa: Economic Commission for Africa.

ECA, 2001, *State of the Environment in Africa*, Addis Ababa: Economic Commission for Africa.

ECA, 2002, *Economic Impact of Environmental Degradation in Southern Africa*, Lusaka: Economic Commission for Africa.

Darkoh. M.B.K., 1998, 'The Nature, Causes and Consequences of Desertification in the Drylands of Africa', *Land Degradation Development*, Vol. 9, pp. 1–20.

Darkoh. M.B.K., 2003, 'Regional Perspectives on Agriculture and Biodiversity in the Drylands of Africa', *Journal of Arid Environments*, Vol. 54, pp. 261–79.

Douglas-Hamilton, I., 1987, 'African Elephants: Population Trends and their Causes', *Oryx*, Vol. 21, pp. 11–23.

Enviro-Protect, 1997, *Illegal Logging and Timber Trade in Cameroon: Background and Consequences. Cut and Run Project*, Vol. 2. Enviro-Protect.

FAO, 1987, *Tree Growing by Rural People*, Rome: Food and Agricultural Organisation.

FAO, 2003, *State of the World Forests*, Rome: Food and Agricultural Organisation.

French, H.F., 1990, *Clearing the Air: A Global Concern*, Worldwatch Paper, No. 94, Washington DC: Worldwatch Institute.

Ghai, D., 1992, 'The Social Dynamics of Environmental Change in Africa', *Whydah, African Academy of Science Newsletter*, Vol. 21(a), pp. 1–8.

Hulme, M. and Kelly, M., 1993, 'Exploring the Links between Desertification and Climate Change', *Environment*, Vol. 35, No. 6, pp. 1–11, 39–45.

Hunter, M.L., 1996, *Fundamentals of Conservation Biology*, Oxford: Blackwell Science.

INCD, 1994, *Elaboration of an International Convention to Combat Desertification in Countries Experiencing Serious Drought and/or Desertification and Particularly in Africa, Final Negotiations, Text of the Convention*, Geneva: UN.

Inyang, E., 2002, *The Effects of Wildlife-human Conflict on Conservation initiatives: A Case of the Banyang-Mbo Wildlife Sanctuary in Southwest Cameroon*, MSc dissertation, Glasgow: University of Strathclyde.

Jordan, C.F., 1987, *Amazonian Rainforests: Disturbance and Recovery*, New York: Springer-Verlag.

Keller, M., Jacob, D.J., Wofsy, S.C., and Harris, R.C., 1991, 'Effects of Tropical Deforestation on Global and Regional Atmospheric Chemistry', *Climatic Change*, Vol. 19, pp. 139–58.

Kuznets, S., 1955, 'Economic Growth and Income Inequality, *American Economic Review*, Vol. 45, pp. 67–98.

Lal, R., 1994, 'Tillage Effects on Soil Degradation, Soil Resilience, Soil Quality, and Sustainability', *Soil Tillage Research*, Vol. 27, pp. 1–8.

Lockwood, M., 2000, Population and Environmental Change: The Case of Africa", in: P. Sarre and J. Blunden, eds., *An Overcrowded World*, pp. 59–107. Oxford: The Open University, Oxford University Press.

Mannion, A.M., 1995, *Agriculture and Environmental Change: Temporal and Spatial Dimensions*, Chichester: John Wiley and Sons Ltd.

Martin, C.,1991, *The Rainforests of West Africa*, Basel: Birkhauser Verlag.

Peters, C.M., 1998, 'Ecological Research for Sustainable Non-wood Forest Product Exploitation: An Overview', in , T.C.H. Sunderland, L.E. Clark and P. Vantomme, eds., *Current Research Issues and Prospects for Conservation and Development*, Rome: FAO.

Postel, S. and Heise, L., 1988, *Reforesting the Earth*, Worldwatch Paper, No. 83, Washington DC: Worldwatch Institute.

Robinson, J.G., 1996, 'Hunting Wildlife in Forest Patches: An Ephemeral Resource', in J. Schelhas and Greenberg, R., eds., *Forest Patches in Tropical Landscapes*, Washington DC: Island Press, pp. 113–30.

Struhsaker, T.T., 1998, *Ecology of an African Rain Forest: Logging in Kibale and the Conflict between Conservation and Exploitation*, Gainsville: The University ofFlorida Press.

Tiffen, M., Mortimore, M., and Gichuki, F., 1994, *More People Less Erosion: Environmental Recovery in Kenya*, Chichester: John Wiley.

UNCED, 1992, *Earth Summit Agenda 21: Programme of Action for Sustainable Development*, New York: UNEP.

UNSO,1992, *Assessment of Desertification and Drought in the Sudano-Sahel Region 1985–1991*, New York: United Nations Sundano-Sahelian Office.

Wilcox, A. and Nzouango, D., 2000, *Bushmeat Extraction Survey within the Banyangi and Mbo Tribes in the Southwest Province of Cameroon*, a final report prepared for the Wildlife Conservation Society/Cameroon Biodiversity Programme.

World Bank, 1992, *The World Development Report 1992*, Washington DC: Oxford University Press.

WWF, 1988, *WWF Year Review 1988*, Gland: WWF International.

Chapter 6

Africa and Global Environmental Problems

Introduction

In the previous chapter, there was detailed analysis of the environmental problems confronting the African continent, otherwise referred to as local and regional environmental problems in our continent. These problems, unlike the global problems we are about to discuss, have yet to receive adequate treatment for a better understanding of their impact on the African continent. Global environmental problems will not be analysed in such depth, because much has already been written on the subject. Global problems refer to widespread or global impacts, irrespective of the geographical areas of origin. Although their causes are largely linked to industrialised countries, and there is a very insignificant contribution from Africa, their impacts do not exclude this continent. Global environmental problems include air pollution, global warming and climate change, ozone layer depletion, acid deposition, nuclear waste and waste treatment.

Air Pollution

Air pollution is of very serious global concern (Figure 6.1); however at the global level, the causes are more closely linked with industrialised countries. However, it should be noted that every day, even in non-industrialised areas, the air is constantly polluted with smoke from kitchens, vehicles and industry, and through slash-and-burn agriculture, and the indiscriminate burning of refuse. The gases contained in the smoke vary with the sources, but carbon dioxide (CO_2) and carbon monoxide (CO) are the most common. Other smoke-borne substances include lead (PB), nitrogen oxide (NO_x), and sulphur dioxide (SO_2).

CO_2 is a greenhouse gas responsible for global warming, which has already been discussed. CO_2 is mainly produced by the incomplete combustion of carbon-containing fuels (coal, oil, charcoal or gas) and incineration of biomass or solid waste (Cunningham et al. 2003). Once it enters the respiratory system, it binds to haemoglobin and interferes with the transport of blood. The results are impaired perception, frequent headaches and drowsiness.

Airborne lead is derived primarily from fuel additives, metal smelters and battery manufacturing plants. But leaded petrol is by far the greatest source in the industrialised world. Direct inhalation of lead can lead to circulatory, reproductive, nervous and kidney damage in adults. Children and foetuses are susceptible to even lower levels of lead. They can face reduced birth weight, impaired mental and neurosensory development, and learning difficulties.

Figure 6.1: 1890 Industrial Model of Smokestacks Pollutes the Atmosphere

Source: Cunningham et al. 2003.

Evaluating the cost of air pollution, as will be seen from the effects of the associated problems, is complicated. For instance, should valuation be limited to the damage of the biophysical environment? Where do the limits lie? Or should it include the secondary health, psychological, social and economic effects suffered by human communities as a result of the pollution? What values can be placed on the effects, which would honestly satisfy each and every affected individual? Must the value placed on these effects be the same as those resulting from natural disasters, or higher?

Serious steps should be taken to arrest this problem, including the education of communities in appropriate waste management methods and appropriate technologies, for example, improved local stoves. Equally, there should be strict enforcement of regulations binding signatory countries to international conventions, which stress the need for radical improvement in industrial processes, with a view to reducing the emission of air polluting substances. There is also a need to transfer cleaner technology from the developed world to the developing nations, which mostly still maintain very old models of technology (Figure 5.7).

Global Warming and Climate Change

Scientific evidence shows that the temperature of the earth is increasing at an un-precedented rate. It is estimated that the global temperature has increased by 7ÚC over the last fifty years. This increase is due to what is referred to as the 'greenhouse effect'. The greenhouse effect is the deflection back to the earth of heat trapped by some pollutants emitted into the atmosphere. These air pollutants, popularly known as greenhouse gases, include carbon dioxide, nitrous oxide, chlorofluorocarbons and methane.

These gases are released from such human activities as the burning of fossil fuels, agriculture, deforestation, and industrial processes, and they vary in their con-tribution to the problem (Figure 6.2).

Figure 6.2: Contribution to Global Warming of Various Types of Human Activity

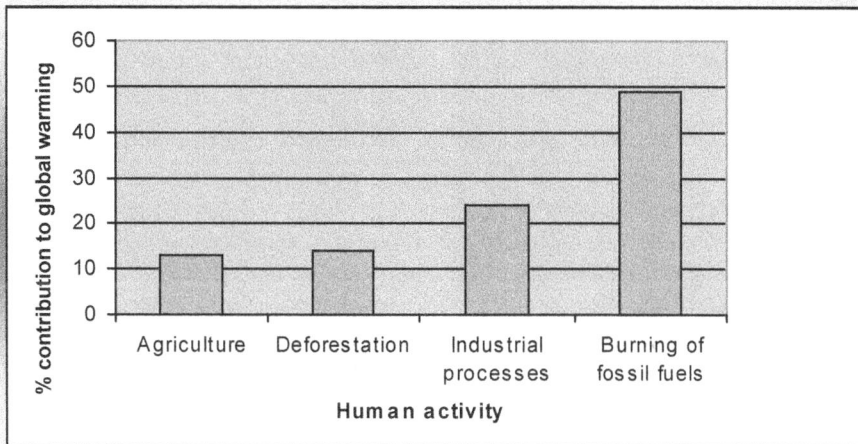

Source: Modified from Cunningham, et al., 2003.

Figure 6.3 shows the relative contribution to the phenomenon by human-induced release of each of the greenhouse gases. The continuous and increasing warming of the earth by these gases is contributing to human-induced climate change. As de-fined in the section on desertification, 'climate change' refers to the short-term climate variability and longer-term climatic trends or shifts caused by natural mecha-nisms or by human activity (Hulme and Kelly 1993). The climate has been changing naturally, constantly but slowly, for hundreds of millennia. As a result of the slow advance of natural processes, the planet has warmed and cooled, passing through ice ages to warm interglacial periods. These gradual transitions, often spanning thou-sands of years, have made it possible for life on earth to adjust relatively smoothly to each new climatic equilibrium. The results have included obvious shifts in the boundaries of ecological communities. Associated human cultures flourished, and, occasionally, disappeared during the transitions (Darkoh 1998). Today the phenom-enon is accelerating due to human activities. This has disastrous effects for life on earth, as it does not allow for any smooth adjustments to new climate regimes.

Figure 6.3: Relative Contributions of Human-induced Greenhouse Gases

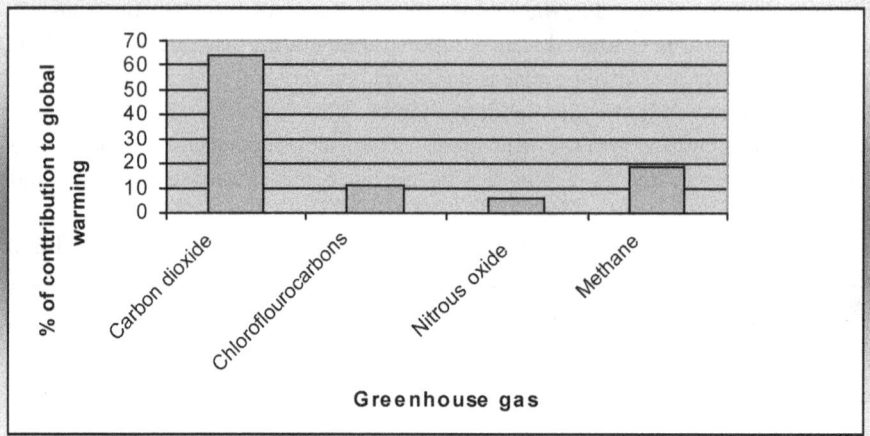

Source: Modified from Cunningham, et al., 2003.

According to the ECA (2002), Africa contributes very little to global climate change, with low carbon dioxide emissions from fossil fuel-use and industrial production in both absolute and per capita terms. Africa accounts for only 2–3 per cent of world's carbon dioxide emissions from energy and industrial sources, and 7 per cent, if emissions from land use (forests) are taken into account. South Africa is by far the largest emitter of carbon dioxide in Africa, responsible for about 39 per cent of the continent's total emissions. South Africa's per capita carbon dioxide emissions (1.88 tonnes) are higher than the global average of 1.13 tonnes per year year. Contrastingly, studies from the period 1990–6 found that Zimbabwe, like other forested countries of Africa, is a 'greenhouse gas sink', as the country's forests are able to absorb a higher quantity of gases than all other sectors emit (ENDA cited in ECA 2002).

Global warming, evidenced in the melting of the polar ice caps, causes an apparent increase in sea levels during the wet periods, resulting in severe flooding of low-lying areas, with the attendant destruction of human life and property. It is also characterised by high incidents of drought during the dry periods, resulting in the destruction of delicate habitats, for example, the drying up of swamps and water bodies, with the concomitant destruction of aquatic life. Droughts, cyclones, floods and bushfires have brought untold hardship to millions of people in southern Africa (ECA 2002).

"'Global warming' is described as 'global warning' by some environmentalists" since it signals catastrophes of global scale, and of serious global concern. Drastic steps must be taken to halt this; otherwise life on earth itself will be at stake. The rate of air pollution and the destruction of the ozone layer must be drastically reduced. This can be achieved mostly through strict implementation of the terms of international conventions that seek to reduce emissions of greenhouse gases, particularly from industrial activities.

Ozone Layer Depletion

Ozone is a gas high up in the atmosphere in the stratum known as the stratosphere. It protects the earth from the ultra violet (strong harmful) rays of the sun. Ozone is constantly formed and destroyed naturally through a series of chemical reactions. But the rate of destruction today is increasing, with the result that the ozone layer or shield over Antarctica is gradually being eroded (a hole was discovered over Antarctica in May 1985). This ozone shield depletion is due to the addition into the atmosphere of chemicals such as chlorofluorocarbons (CFCs), emitted from products such as aerosols, fridges and fire extinguishers.

CFCs are resistant to chemical break-down in the troposphere. Therefore they can move slowly upward into the stratosphere where they are carried by strong winds across oceans and continents (Westman 1985). At that high level, they finally find themselves in the appropriate, very cold, region, where they encounter frozen particles, with which they react chemically, releasing the chlorine atoms they contain. One atom of chlorine can destroy 100,000 molecules of ozone (Mader 1990; Baird 1999). This is possible because the chlorine atom is not involved in the chemical reaction with ozone and so does not get destroyed, but rather acts as a catalyst.

In order to reduce, or stop, the rapid destruction of the ozone layer, individuals would have to refrain from using products that contain CFCs. Also, there should be a mechanism for ensuring that manufacturers are given an incentive to use substitutes that are not dangerous to ozone. This could be either through the implementation of subsidies for substitutes to ozone depleting substances, or through stricter enforcement of the regulations of the convention that aims to cut down, and eventually ban, the use of such substances.

Acid Deposition

This phenomenon, like air pollution, is largely associated with highly industrialised regions, characterised by high levels of coal burning (Westman 1985). As a result of the increasing tide of industrial activities, large quantities of nitrogen oxide and sulphur are emitted into the air. They undergo chemical reactions and become acids as they are absorbed by the moisture. These substances finally return to the earth in the form of rain, fog or dust, causing damage to forests, buildings and other materials. They also result in the acidification of soils and aquatic ecosystems: streams, rivers and lakes. The end result is unprecedented crop failure and massive losses of aquatic life.

The only possible and visible solution to the problem of acid deposition is the stringent implementation of the laws and regulations, which have been set by the international convention aimed at abating the problem. They call for radical reforms in industrial processes to curb air pollution, enforceable with strict sanctions against violating nations. This could involve the application of the 'polluter pays' principle (PPP), which requires that the polluting nations are asked to bear the cost of the air pollution problem. It is important to note that the PPP is not only applicable and limited to pollution: it applies to all environmental problems. It is a means of enhancing the capacity of governments to deal with environmental and development issues in a cost effective manner, promoting technological innovation, influencing consumption and production patterns, and providing an important source of funding (Panayotou cited in ECA 2002).

Nuclear Wastes

The discovery in 1942 of the use of nuclear energy by the Italian physicist, Enrico Fermi, brought in what is known as the nuclear age. Nuclear energy is produced by splitting uranium atoms. This new technology has proved important in the areas of warfare, medicine and electricity.

Many developed countries, and some developing ones, have well established nuclear plants for the production of nuclear weapons, or the generation of nuclear power, which is considered to be very cheap. However, there are strong moves to prevent any more countries from developing nuclear technology. This is not only because of the fear of proliferation of 'weapons of mass destruction', as nuclear weapons are now branded, but also because of the serious problem of nuclear waste.

Nuclear wastes are highly hazardous and causes serious health problems, such as body tissue damage, cancer, degenerative diseases, for example, cataracts, mental retardation, genetic disorders and weakened immune systems. The materials are radioactive, and can spread quickly through the environment by means of air and water. Materials from a nuclear plant can affect organisms thousands of miles away. An atmospheric nuclear bomb testing created radioactive 'fallout' that spread around the globe (Lenssen 1991).

The problem of nuclear wastes accumulation is a great concern. The waste created since the beginning of the nuclear age is yet to be effectively managed. Packaging and burying wastes deep in the earth was thought to be the best option. But it is now clear that there is no guarantee that the materials will remain permanently sealed off from the biosphere. There is evidence of leaking, which could result from tectonic crushing or chemical bursting of the containers in which the wastes are stored, as well as from the corrosive and contamination-spreading action of groundwater.

The technology of packaging wastes minimises the risks to the present generation. But its use is controversial because of the danger it poses to future generations, given that the materials remain radioactive for hundreds of thousands, or millions of years. Several other methods have been proposed, some of which have been discarded, due to their associated risks, while others are being studied. However, none of the current methods have the potential to address the problem of radioactivity of nuclear waste. So far, only the natural decay process is known to diminish it; however, unfortunately, this takes a long time (Baird 1999). This suggests that the nuclear business is risky and should not be encouraged until there is more advanced technology to address the question of waste management.

Treatment of Wastes and Pollutants

All along, we have been stressing the need for the prevention of pollution. But what if a level has been reached where prevention is no longer the only solution? How can we address situations where soils, sediments and water bodies have already been loaded with wastes or chemical pollutants? Different methods are used to address the problems of the accumulation of solid wastes and the contamination of soils and sediments by anthropogenic (man-made) chemicals.

Solid Wastes

The production of solid wastes in most big cities of the world is alarming. In some cities at least hundreds of tonnes of wastes are produced every day, from manufacturing, packaging, construction and demolition work. These are normally dumped in specific sites, where municipal council trucks go round to collect them.

Some of the wastes are disposed of as landfill, the cheapest method. We have already identified the problem associated with landfill: contamination of groundwater in the course of waste decomposition. But there are improved landfills in which the holes are lined with thick plastics to prevent the leachate (liquid that settles at the bottom of decaying materials) seeping into the groundwater. In a well organised system, landfills are not used for hazardous wastes, and are located where they have minimal impact on the environment.

Other solid wastes, particularly organic matter, are eliminated by incineration. A good incinerator is capable of reducing wastes into their simplest forms, with reduced gaseous emissions. However, the incineration of hazardous wastes is complicated, fraught with the possibilities of more serious effects. It requires more advanced technology. Baird (1999) described three types of incinerators, which have been developed to address this situation: molten salt combustion, fluidised incinerator and plasma incinerator, which make use of very high temperatures and other substances. Additionally, supercritical fluids are being employed as a more modern alternative to traditional incinerators.

Recycling is also used as a method of managing solid wastes. Four 'R's constitute the philosophy of waste management: *reduce* the amount of materials used in the manufacture of products, *reuse* the materials once they have been made into products, *recycle* the materials by fabricating them into new products, and *recover* the energy content of the materials if they cannot be reused or recycled (Baird 1999). The purpose of recycling, or more generally waste management, is to reduce the amount of waste produced and to conserve the natural resources from which products are made.

Chemical Pollutants

One of the results of pollution is the contamination of soils, including sediments, and water. Some of the pollutants are so toxic that they destroy plants, including crops, or, at best, inhibit their growth. Other pollutants are less toxic, and taken up by the roots and become part of the food chain, where they cause health and reproductive problems. Two important technologies have been developed to remove pollutants from soils, sediments and water. These are: bioremediation and phyto-remediation. Bioremediation describes the use of micro-organisms, such as bacteria and fungi, to degrade or break down wastes and pollutants, as they utilise waste as food substances. Nitrogen-containing fertilizers are applied to the contaminated soils and sediments to stimulate the rapid growth of micro-organisms that facilitate the biodegradation process. It should, however, be noted that some pollutants and wastes are not biodegradable; these are known as recalcitrant.

Phytoremediation is the use of plants for the decontamination of soils and sediments that have been loaded with heavy metals, such as lead, or organic pollutants. This is an attractive technique, because metals are often difficult to extract using other technologies due to their low concentrations (Baird 1999). When pollutants are absorbed by plant roots, they may be incorporated into the biomass, while a small quantity are emitted into the atmosphere. Some are broken down by oxygen and enzymes are released by the roots; the enzymes of fungi and microbes are released from the plants. Once the plants have done the work of decontamination, they are harvested and burnt in time to prevent recontamination through leaf-fall.

Conclusion

This chapter has concentrated on the analysis of global environmental problems: air pollution, global warming and climate change, ozone layer depletion and acid deposition. It has highlighted their causes and impacts and proposed possible practical control measures. The chapter also touched on issues of nuclear waste, describing methods for treating not only nuclear waste, but also solid and chemical waste.

Revision Questions

1. With at least three examples of environmental problems, distinguish between the local and the global problems.
2. Name three global environmental problems, analyse their causes, and propose solutions.

Critical Thinking Questions

1. What is the correlation between deforestation and global warming?
2. Air pollution and acid deposition are problems associated more with industrialised nations. Why should an African worry about these problems?
3. Can the saying 'Global warming is global warning' be justified?
4. Nuclear technology is both a blessing and a curse to the world. Discuss.

References

Baird, C., 1999, *Environmental Chemistry*, New York: W.H. Freeman and Company.

Cunningham, W.P., Saigo, B.W. and Cunningham, M.A., 2003, *Environmental Science: A Global Concern*, New York: McGraw-Hill.

Darkoh, M.B.K., 1998, 'The Nature, Causes and Consequences of Desertification in the Drylands of Africa', *Land Degradation Development*, Vol. 9, pp. 1–20.

ECA, 2002, *Economic Impact of Environmental Degradation in Southern Africa*, Lusaka: Economic Commission for Africa.

Hulme, M. and Kelly, M., 1993, 'Exploring the Links between Desertification and Climate Change', in *Environment*, Vol. 35, No. 6, pp. 1–11, 39–45.

Lenssen, N., 1991, 'Nuclear Waste: The Problem that Won't Go Away', in *Worldwatch Paper 1.6*, Washington DC: Worldwatch Institute.

Mader, S.S., 1990, *Biology*, Dubuque: Wm. C. Brown Publishers.

Snyder, ed., *The Biosphere Catalogue*, Texas and London: Synergetic Press Inc.

Westman, W.E., 1985, 'Global Pollution Analysis, Monitoring and Control', in T.P.

PART II

SELECTED TOOLS FOR
ENVIRONMENTAL MANAGEMENT

Chapter 7

Conservation and Natural Resource Management

Introduction

The history of conservation and natural resource utilisation is as old as the human race. Interestingly, it is characterised both by the nullification of the user rights of communities, and by devolution of power to the communities. On the one hand, nullification of user rights was a measure adopted by many nations to control access to natural resources, ostensibly for the purpose of management. But it was obviously also a strategy for the monopoly of substantial amounts of the benefit accruing from resource exploitation. On the other hand, devolution of power to the communities was necessitated by the realisation of some governments of their inability to effectively and efficiently control resource exploitation, and of the costs and risks associated with the responsibility. This highlights the significance of the generally held view that the success of natural resource management (NRM) depends on the involvement and active participation of communities that traditionally or naturally have rights of access and use of resources to satisfy their basic necessities, hence the term 'community-based natural resource management'.

Today, natural resource management has become important, especially to NGOs around the globe. The role of these NGOs is to encourage the devolution of power to rural communities by reluctant governments, and in building the capacities of communities. But despite their efforts, problems still abound around the management of natural resources in many parts of the world. This chapter highlights some of the problems as it attempts some definitions, classifications and arguments associated with natural resource utilisation.

Definitions

Natural resources are all the natural items (organic or inorganic) that provide our sources of food, medicine, building and ornamental materials necessary for our daily lives and pleasures. This definition, though simplistic, especially from an ecological perspective, is apt within the context of NRM. Although non-human living organisms utilise a wide range of resources, the combined ecological consequences of their utilisation are minor compared with human use.

Debates in conservation and NRM circles emphasise that a world without humans would leave the earth and its resources in their usual dynamic but self-sustaining state. We humans are by nature omnivorous and insatiable. With our rapidly increasing populations and growing demands, humans present a mounting threat to the earth and its limited resources. Fortunately, as humans are also, relatively, the most intelligent living creatures on the planet, we are capable of managing these resources.

Conservation and NRM are synonymous, both referring to the utilisation of natural resources in ways that allow the present generation to satisfy their needs, and make it possible for future generations to satisfy their own needs (Begon et al. 1996; Neba, 2005). Conservation, or NRM, is the sustainable management or conservation of natural resources. The process involves preventing resources from rapid exploitation and pollution. Resource depletion results from over-exploitation and/or from poor harvesting. Over-exploitation is closely linked to what may be described as economic warfare, characteristic of open-access regimes, which, at the micro-level, force individuals to increase their rate of exploitation in order to maximise their profits (Hardin 1998). Poor harvesting is a feature of societies that still employ unsound exploitation techniques, due to the absence of modern or improved alternatives.

Economic warfare, at the macro-level, encompasses three main types of continuous and never-ending struggle: 1) the struggle by developed countries to maintain their position on the top rung of the economic ladder in order to continue to enjoy their dignity as super powers; 2) the struggle by developing countries to also reach the top rungs of the global economic ladder in order to gain recognition as super powers; and 3) the struggle by the undeveloped countries to get out of the abyss of poverty in order to seek ways of developing and liberating themselves from characteristic aid-dependent mentalities.

Let us turn to the question of resource pollution. This can be described as the unavoidable economic by-product of using chemicals, or of embarking on genetic manipulation or engineering without proper care, but with the primary aim of achieving maximum resource yield. Often, polluting chemicals are manufactured. They are readily preferred by the poor consumer nations, because they are cheap. However, the lack of political will in the application of sound technologies by the manufacturing nations, the weakness of international legislation to control manufacturing, marketing and use of the chemicals, and the low awareness levels of a majority of the poor countries about the real human impacts of these chemicals contribute to their continued manufacture and use.

As intelligent humans, it is incumbent on us to take precautions as we observe the obvious signs of rapid resource depletion and pollution. Either of these situations can eventually lead to an irreversible phenomenon known as extinction (Figure 7.1). Rapid depletion results from the fact that when resources are over-exploited, very little or no room is allowed for their regeneration. Pollution either destroys resources directly, or suppresses their ability to function properly and regenerate.

Regeneration, however, is not characteristic of all natural resources. We shall now consider their classification in order to establish where this term actually applies.

Figure 7.1: Rapid Exploitation and Pollution of Natural Resources Lead to Extinction

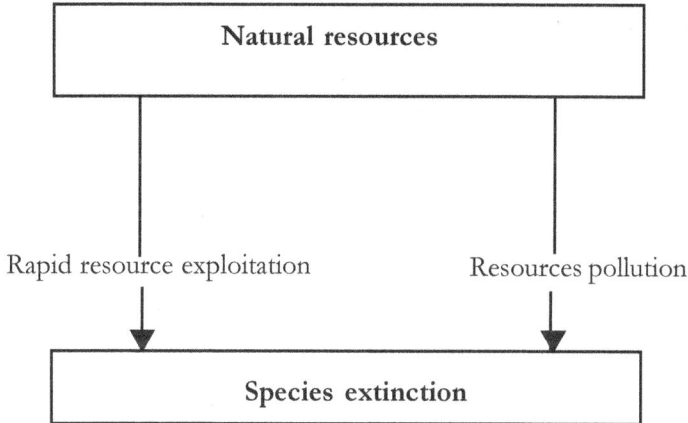

```
┌─────────────────────────────────────────────────────┐
│                 Natural resources                     │
└─────────────────────────────────────────────────────┘
                         │
                         │
  Rapid resource exploitation        Resources pollution
                │                              │
                ▼                              ▼
┌─────────────────────────────────────────────────────┐
│                 Species  extinction                   │
└─────────────────────────────────────────────────────┘
```

Classification of Natural Resources

We are familiar with the classification of resources as biotic and abiotic. This classification is common to all biological texts. It simply illustrates whether a resource is composed of living or non-living materials, and whether the resource itself is living or not. In NRM, this classification is extended to include what these resources are capable or not capable of: what is their economic interest to us. Our concern is whether as we continue to exploit these resources, whether they are able to regenerate, or not, hence their broad classification as renewable and non-renewable. Renewable resources are those whose stocks are capable of being decreased, through intra- and inter-specific interactions and human interference. They can increase within the carrying capacity of the ecosystems in which they are found, as a normal or compensatory response made possible by their potential to regenerate or renew their stocks, such as a forest or a population of animals, if and when the impacts of competition or exploitation are within tolerable thresholds.

It should also be noted that resources exhibiting continuous flow through time, such as the energy from the sun, and the waves or tides, are also considered renewable (Pearce and Turner 1990). However, we have no influence or control over these continuous flow resources. Their consideration in NRM is not for their management per se, but for their potential uses as alternative energy. Similarly, non-renewable resources such as mineral resources, including fossil fuels, cannot regenerate their stocks, and deserve proper care in harvesting them. For instance, once an oil well or a coal mine has been exhausted, it cannot be replenished.

Renewable and non-renewable resources may both play equally important roles in our subsistence and economic lives. However they pose completely different problems and challenges to our management strategies. It goes without saying that the challenge to managing non-renewable resources is far greater to manage renewable resources. Figure 7.2 provides an illustration of how to distinguish between renewable and non-renewable resources, as well as some strategic management options that could be applied to each of these classes.

Figure 7.2: Some Options for Managing Renewable and Non-renewable Resources

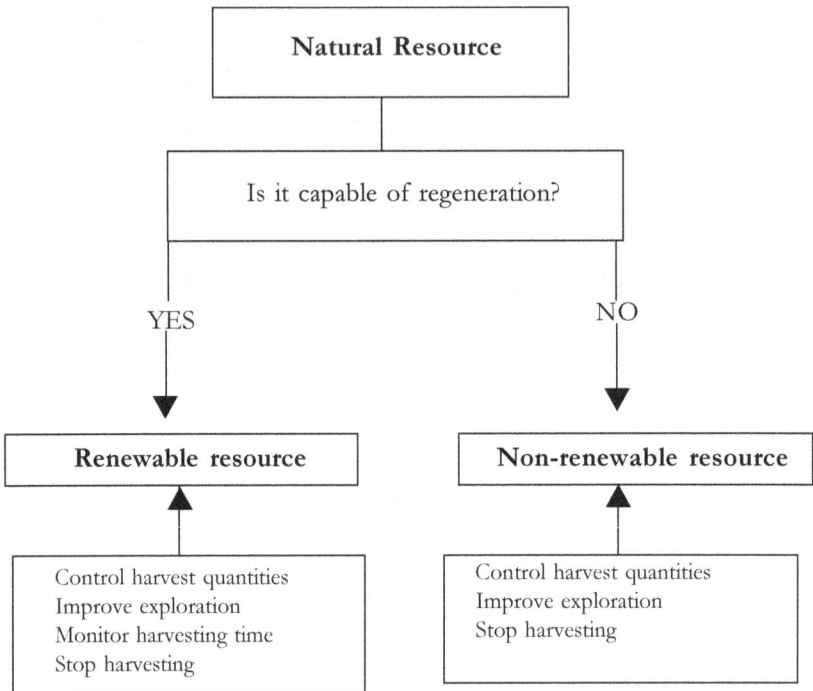

Conservation, or NRM, should, as a matter of principle, focus mostly on non-renewable resources, while paying proper attention to renewable resources, because they hold more promise for sustainability, if properly managed. Although the latter have the potential to regenerate, human exploitation rates may far exceed their regeneration rates. Exploitation techniques employed may be too incompatible with the facilitation of their regenerative capacities. *De facto*, the over-exploitation or the application of poor exploitation techniques could render renewable resources 'non-renewable', hence the imperative to promote the sustainable utilisation of renewable resources. To guarantee this requires not only the application of legislation and development of regulatory mechanisms, but also the implementation of studies to properly understand their ecology, the promotion of capacity building and institutional strengthening efforts geared towards NRM at local, national and regional levels.

This century has seen an explosion in capacity building and institutional strengthening on the part of governments and NGOs, as well as important studies of a broad range of natural resources. One result has been their classification into wood forest products, further divided into timber and non-timber forest products, and non-wood forest products, which include animals of various types (Figure 7.3).

Figure 7.3: Classification of Renewable Forest Resources

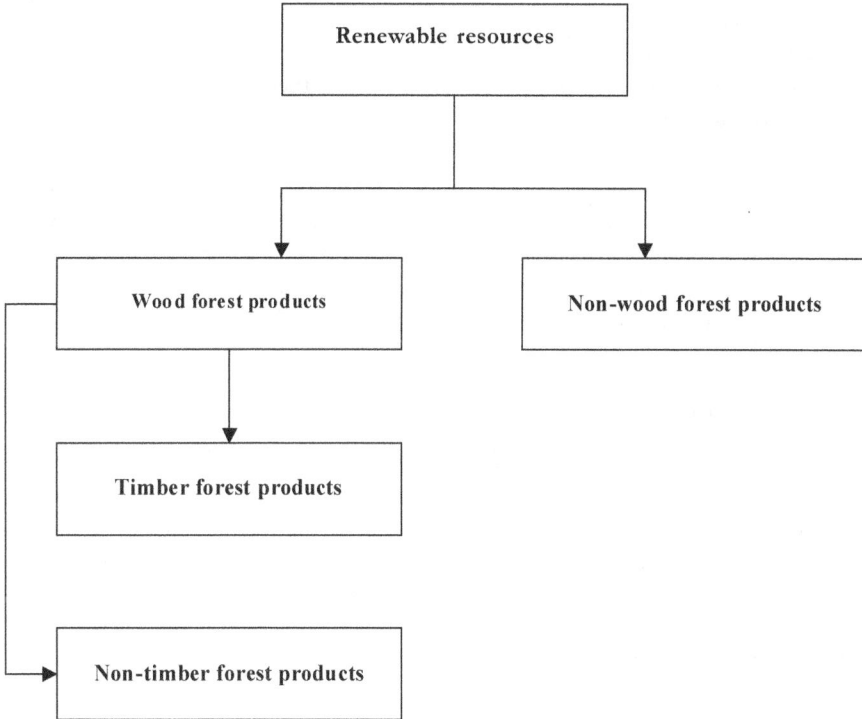

Because of the increasing rate of species extinction, due to human activity, there is now global emphasis on conservation, defined, simply, as the sustainable use of natural resources. Conservation is also defined as the management of the biosphere in order that it may yield the greatest sustainable benefit to the present generations, while maintaining its potential to meet the needs and aspirations of future generations.

A sustainable process or condition can be maintained indefinitely, without progressive diminution of valued qualities, inside or outside the system in which the process operates or condition prevails (Holden et al. 1995). Sustainable development can be defined as any type of development that yields long-term benefits with minimal or no impact on the environment, which otherwise would reverse the situation. *Our Common Future*, the 1987 report of the World Commission on Environment and Development, chaired by Norwegian Prime Minister Gro Harlem Bruntland (hence the report's popular title, the Bruntland Report), defined sustainable development as development that satisfies the needs of the present generations without compromising those of future generations (Cunningham et al 2003).

Conservation and sustainable development are increasingly treated as two sides of the same coin. In actuality, conservation per se may be repellent to some communities while development is a readily welcomed idea. This is so because the communities may not see the benefits of conservation as clearly as those of development. Conservation may even be seen as a means of controlling or prohibiting the use of some natural resources that have long been part of the economy. Whereas development is readily considered as means to improve the welfare and living standards of the population.

It is in view of these misconceptions about conservation that some projects now include aspects of development in their plans, at least to demonstrate their intention to ensure long-term benefits to local populations, who are not sufficiently patient to wait for conservation benefits that often take time to accrue. Some projects are designed as integrated conservation and development projects (ICDPs). This is obviously to ensure the adoption by the rural communities of sustainable development ideals, which do not undermine conservation needs.

Arguments for Conservation

As already indicated, it is often difficult to convince rural populations that conservation is serving their own best interests. It may be advisable to state how natural resources benefit these populations, and how by not using these resources properly could lead to their extinction. Taylor et al. (1997) identified three principal areas in which arguments for conservation can be made: economics, ecology and ethics.

Economic Perspective

The *economic* perspective considers the direct and potential values of resources. Many of these resources contribute significantly to the rural economy, as food, building materials, medicines and tourist attractions. Many more have the potential to do so. The over-exploitation of these resources, at levels that may lead to their extinction, could mean huge losses to the populations.

Ecological Perspective

Some resources play important ecological roles that contribute to the distribution and abundance of other resources or species. For example, some animals act as agents of dispersal or pollination for a wide variety of plants. While others act as key predators that help in the maintenance of prey populations in order to reduce the negative effects of intra-specific competition. Others still may also act as principal prey that feed the populations of a number of predators. The seeds of certain plants can only germinate after they pass through the guts of certain animals, e.g., elephants. Leguminous plants are also important in enriching the soil.

Ethical Perspective

Ethically, every species has a value in its own right, and, therefore, the right to live. It is true that the values of some species are difficult to appreciate. This could make any support for their conservation sound ridiculous. But our inability to appreciate their value is not the fault of the species. It is largely due to the fact that our knowledge of the species is generally rather embryonic.

Considering our limited knowledge, we may never know whether the destruction of a species, today seen as having no value, could result in the disruption of an ecosystem. It is also important to note that the loss of such species means the loss of genetic diversity. With improvements in knowledge and technology, these species may come to be seen as indispensable to genetic engineering or medical advances.

Cultural Perspective

Though not universally applicable or scientifically explicable, it might be necessary to add a fourth perspective to the arguments for conservation, namely, culture. In some parts of Africa (and this is true also for other parts of the world) there is an intricate relationship between nature and culture. Many traditional cultures are characterised by secret societies that depend on specific natural resources for their rites and ceremonies (Figure 7.4). It is easy to guess what would happen to these societies should resources become depleted. Similarly, in some of the cultures, certain individuals, families or clans maintain close spiritual (or supernatural) links with particular animal species, referred to as totems.

Figure 7.4: Traditional Ceremonies Depend on Natural Resources

Photograph taken at Tangang village in the southwest region of Cameroon.

Family or clan totems are believed to be responsible for the protection and general welfare of social units. Their destruction is believed to have serious repercussions for the social unit. Individual totems are acquired for self-protection but mostly as a source of supernatural powers against enemies. It is believed that the death of a totem, often as a result of poaching, could result in the death of the individual associated with it, except in a case where there is a powerful traditional doctor who could intervene immediately to link the individual to a new totem of the same or a different species.

In discussions about the importance of conservation with people from strong cultural backgrounds, it might be instructive to use the cultural basis as an entry point, before moving on to the economic, ecological and ethical considerations. Many species and habitats are known to have survived to this day as a result of community adherence to traditional beliefs and taboos. Some species, including certain totems, are regarded as taboo-animals. They must not be harmed at all, or eaten, by certain groups or individuals.

Conservation Strategies

Increased knowledge of and concern for environmental problems have motivated individuals, NGOs, and private and public institutions to seriously engage in conservation in various ways and at different levels. There are two approaches to conservation, *in-situ* and *ex-situ* conservation, irrespective of its methods, and the level at which efforts is directed.

In Situ Conservation

In situ conservation describes a situation where the conservation initiative is implemented directly at the site where it is intended. This comprises the identification, designation and management of natural areas to give them various levels of protection, as defined by conservation objectives. These objectives could stress the protection of whole habitats, including all the species therein, or of a specified number of species they contain. From these broad definitions, designated areas are classified as fully protected or partially protected.

Fully protected areas include, in order of importance, world heritage sites and national parks. These are areas where human activities, such as hunting, harvesting or the collection of natural resources and agricultural activities, are strictly prohibited. Partially protected areas include reserves and sanctuaries. Reserves are classified as production and protection reserves. In production reserves, individuals with permits are allowed to use resources within specified limits, whereas in protection reserves, no one is allowed to use resources until such a time as when the reserves change their status to permit human use. In the case of sanctuaries, individuals, again with permits (or user rights), are allowed to use all but those resources that they are created to protect, or are protected throughout the national territory.

It is important to observe that protected areas serve as safe havens where species breed and multiply freely, without human interference. Naturally, these species spill over into areas outside the boundaries of the protected areas, where local communities are free to harvest them with fewer restrictions, taking advantage of user rights; but in accordance with the law. In such situations, it is important that everything possible is done to improve harvesting methods, and to control the harvesting effort. The cruder the methods, the higher is the possibility that resources will be destroyed in the process of harvesting. The bigger the harvesting effort, the greater is the proportion of stock to be harvested. Intensification of harvesting could place it far beyond the sustainable levels of each given resource: above levels that leave room for the regeneration of the resource. But controlling the harvesting effort in such open-access regimes is difficult to achieve, as each individual strives to maximise profits from the common pool. Fortunately, devolution of power and management responsibility to local communities have proved to be effective means of redress. However, for the arrangement to be successful, appropriate tools and mechanisms, which guarantee sustainability, must be implemented.

The effective management of any protected area depends on a management plan. This document gives a clear description of the biophysical aspects of the area, the kinds of activities and pressures that constitute a problem, and the management options that could be applied to zones in the area. Drawing up a management plan involves several years of basic and applied research within biological and socio-economic domains. It involves consultations with all stakeholders, including the local communities, local organisations, and relevant government institutions. It should be stressed that a good management plan depends on both expertise and indigenous knowledge of the area concerned.

Ex-Situ Conservation

Now let us consider the next conservation approach: *ex-situ conservation*. This is a process by which individuals, organisations or institutions are engaged in conservation-oriented activities outside the intended conservation areas, with the ultimate aim of promoting conservation of the intended areas. This may require the application of high skills and sophisticated technology to activities such as captive breeding and development of seed banks. Captive breeding programmes are commonly carried out in well developed zoos. This effort has saved many animal species from becoming extinct. Endangered animals are bred in captivity, and are reintroduced into the deprived wild areas where they once roamed.

The development of seed banks is another type of *ex-situ* conservation. A huge collection of seeds of endangered plant species is kept under conditions favouring prolonged storage. The seeds are used for breeding at much later dates in areas where species are either critically endangered or already extinct. The development of sperm banks may be included in this category. This requires at least one living and healthy female species on which to conduct artificial insemination or implantation, depending on how critical the situation is, with regard to the species concerned.

This is very risky, as the result may not lead to the correct level of captive breeding, due to factors ranging from success with the technique to survival of the female species.

Gender Considerations in Natural Resource Management

Gender issues are now widely and frequently discussed and debated at local, national and international levels, due to the recognition of the continued marginalisation of women in many parts of the world. In developed nations, women take part in policy and decision making, and participate in a wide range of cultural events, also in political, economic and environmental activities, at all levels. This is rarely the situation in a majority of developing countries. However, there is now growing interest in the active role of women in development, as can be seen, for example, from the massive participation of women at the conferences in Dakar in 1994 and Bejing in 1995 (Burnley 1999). The exclusion of women from, for instance, environmental management, can have negative impacts at both household and community levels (Commonwealth Secretariat 1996).

The interaction of women with the environment is no less important than male interaction, in terms, particularly, of their effects both on the environment and the human community. In rural areas, women's livelihoods are intimately linked to natural resources. Women are the primary exploiters of non-timber forest products (NTFPs) for both domestic consumption and income generation. In some forest areas of Cameroon and Nigeria, such products include *Invingia gabonensis*, *Gnetum africanum* and *Ricinodendron heudelotin*. Woman are engaged in trading these and other non-timber forest products such as game meat (Figure 2.8), which is also sold by some women as pepper soup (Figure 2.14). The continuing increasing economic values of these products is a function of increasing human populations, and of the rising scarcity and demands for the products.

Rural women spend considerable time processing and transforming raw materials or products like cassava and oil palm nut, into widely consumed and highly priced foodstuffs. In their product processing and transformation role, women are also generators of domestic waste, the disposal of which can have serious implications for both environmental and human health. Furthermore, in many rural communities, women play a major role in the burning of farm plots, and in the cultivation and care of food crops. They are also largely responsible for the collection of fuelwood, the fetching of water and the cooking of meals for the household. In these roles, women are both agents and victims of air and water pollution, and of soil degradation or improvement.

In their more traditional roles as home managers, women are frequently exposed to household chemicals in various products. Their bodies absorb pollutants and toxins, which, in reproductive terms, are passed on to the next generation. Perhaps the most extraordinary feature is that women are also engaged in the collection of waste items for recycling purposes. See for example, the woman depicted in Figure 7.5. This effort brings women reasonable cash, and helps clean the environment where the collection is done. Mostly children from poor homes collect waste items, which they supply to women during their routine trips, to raise cash for their own needs.

Figure 7.5: Women Interact with Different Natural Resources in Various Ways

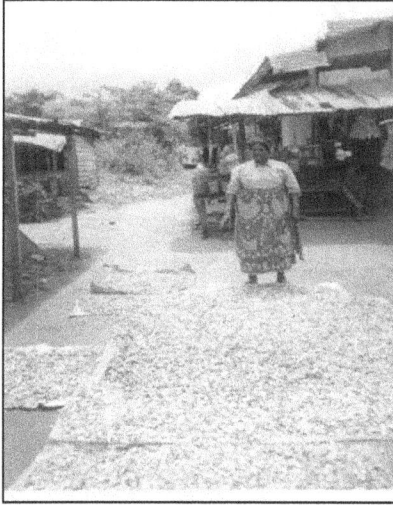

Photograph taken at Nguti town in the southwest region of Cameroon.

By interacting with natural resources in such an intimate fashion, women have, over the centuries, accumulated a stock of knowledge about the local environment, nature and the durability of land, water and other natural resources, besides the medicinal and other uses of these resources (Commonwealth Secretariat 1996). Their indigenous knowledge, together with their multiple roles, can be usefully exploited to give meaning and relevance to environmental education. This could serve as a starting point in the conception of informal and non-formal environmental education programmes. By implication, women should be considered a very important target in the design, development, and implementation of such programmes, particularly because of their additional role of caring for and raising children. In this way, environmental education can also help clarify the roles of women in community natural resource management and be used as a tool for women's empowerment. However, it should be recognised that if women are required to participate in community meetings and workshops, then the time that is required should be factored into their overall commitments (Commonwealth Secretariat 1996).

Women interact with natural resources to obtain products for domestic consumption and income. Unfortunately, until recently, little market attention has been paid to many of these products, possibly because men, who have taken a leading role in many promotional activities, such as search for markets, were not traditionally engaged in their harvesting, collection and marketing. Consequently, the harvesting and gathering methods of NTFPs have largely remained underdeveloped, and, as a result, large-scale processing is absent, and low quality products are common.

Box 7.1: Women are also Engaged in Environmental Sanitation
through Recycling Efforts

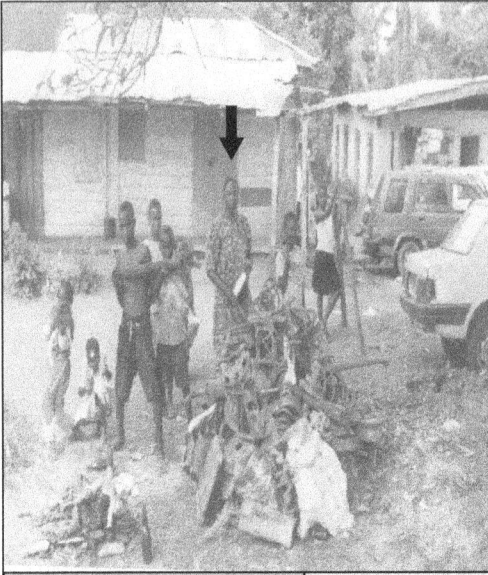

This Cameroonian woman, Fansi Perpetua Chengwe (follow arrow), is a pioneer in the business of collecting waste items for the purpose of recycling. She makes regular trips to surrounding towns, some two hours drive away, and collects old roofing sheets, old rubber shoes and old iron rods which she sells to middlemen in Kumba, her base. The men supply these items to different recycling industries in the country. The old rubber shoes are supplied to Cisplast and the old iron rods to Foukou (both recycling industries in Douala) while the old roofing sheets are sold locally for local production of cooking pots. She has been in this business for almost ten years now. Although she does this primarily to generate income for herself, she is very conscious of the fact that she is helping to clean the environments of the communities where she operates. She is beginning to see some positive changes in the habits of people in the communities. For instance, in the past the people dumped all sorts of rubbish (including kitchen refuse) together, which made what she collected usually so nasty; but now they are beginning to separate the wastes and keep what she would come back for in much tidier conditions.

However, she continues to take precautionary health measures, as she does her business, knowing that no matter how neatly kept the waste items are she will still get in contact with disease germs through them. She is proud to note that the environments in the communities where she operates are taking on a better look due to her activities.

The implication is that very little value is added to these products before they are marketed (Nkwatoh 2000), thereby contributing to low market prices, compared with those of products harvested and marketed by men. This is another important subject of women's empowerment. Moreover, developing the potentials of a wide variety of these products to yield more sustainable income earnings for rural communities would guarantee the conservation of tropical forests (Ndoye et al. 1999). The logic is that a significant increase in the contribution of these products to the rural economy, and, therefore, to rural development could result in a corresponding reduction in the incentive for local communities' support of logging operations. Reaching this stage also depends on the local communities' appreciation of the value of a standing forest in terms of its important economic and environmental functions (Nkwatoh 2000). This points to the need for environmental education.

There is a widespread misconception that the exploitation of NTFPs has no ecological consequences. Since a forest exploited for its non-wood resources, unlike a logged-over forest, maintains an appearance of being undisturbed, as the impacts are subtle and hard to observe to the untrained eye, there is the assumption that these resources can be harvested repeatedly, year after year, on a sustainable basis (Peters 1999). It is argued that even when the harvesting of some NTFPs does not involve damage to, or killing of, the stock, as is the case with the collection of the fruits of bush mango and *njansang*, collection in commercial quantities spells severe competition with frugivores (some of which may also be herbivores) that are naturally forced to increase their foraging to obtain sufficient food to sustain their lives.

What are the consequence? First, the number of seeds actually left behind from the combined activities of commercial collection and foraging is too small to guarantee any recruitment into the population. Second, foragers that could easily switch to herbivory now increase their attack on the vegetation, including what remains as the seedlings of target NTFPs that managed to survive into the germination stage) that would have been somehow spared during that period.

Conclusion

The earth is blessed with both renewable and non-renewable natural resources that serve as insurance for the continuous improvement of the general welfare of humankind. Classification of resources as renewable means that they have the potential to regenerate and increase within the carrying capacity of their various ecosystems. But it is also clear that they can be rendered non-renewable by the human tendency to carry out exploitation at rates higher than their regeneration rates, and to apply exploitation techniques incompatible with their regenerative capacities.

The current global trend shows that natural resources are under increasingly serious stress due to a combination of a number of underlying factors, including poverty, a lack of political will and struggle for economic power. Non-renewable resources present enormous challenges to NRM efforts. The situation would be made even more complex if renewable resources were allowed to degenerate into that category.

There are fundamental arguments that can be advanced for the conservation and sustainable management of natural resources, and well established strategies to achieving conservation objectives. However, arguments and strategies, though useful in education terms and conservation, are not in themselves sufficient to bring about the desired change. For instance, it is increasingly obvious that effective and efficient natural resource management can be achieved only through the involvement and active participation of communities with traditional rights of access to the utilisation of resources to meet their basic requirements. The role of women in natural resource degradation and management must be given utmost importance.

Revision Questions

1. With concrete examples, define natural resources.
2. With examples, distinguish between renewable and non-renewable resources.
3. Conservation and natural resource management are used interchangeably. Define and state the objectives of this practice.
4. Discuss the three basic struggles that economic warfare promotes at the macro-level, which create an impact on the environment.
5. Advance and discuss three arguments in favour of conservation and natural resource management.
6. With illustrative examples, distinguish between *in situ* and *ex situ* conservation.

Critical Thinking Questions

1. Conservation is a determinant of sustainable development. Discuss.
2. The belief in totems offers great opportunities in the conservation of the African environment. Discuss.
3. The effective management of any protected area depends on a management plan. Discuss how this can help in the management of a protected area of your choice.

References

Begon, M., Harper, J.L. and Townsend, C.R., 1996, *Ecology*, London: Blackwell Science.

Burnley, G.E., 1999, 'The Role of Women in the Promotion of Forest Products', in T.C.H. Sunderland, L.E., Clark and P., Vantomme, eds., *Current Research Issues and Prospects for Conservation and Development*, Rome: Food and Agricultural Organisation.

Commonwealth Secretariat, 1996, *Women and Natural Resource Management: An Overview of a Pan-commonwealth Training Manual*, London: Commonwealth Secretariat.

Cunningham, W.P. and Saigo, B.W., 2003, *Environmental Science: A Global Concern*, New York: McGraw-Hill.

Hardin, G., 1968, 'The Tragedy of the Commons', *Science*, Vol. 162, pp. 1243–48.

Holden, P. J., Daily, G.C. and Ehrlich, P.R., 1995, 'The Meaning of Sustainability: Biophysical Aspects', in M. Munasinghe and W. Shearer, eds., *Defining and Measuring Sustainability: Biogeophysical Foundations*, Washington DC: The United Nations University and the World Bank.

Inyang, E., 1996, 'Community Based Natural Resource Management', *Jordanhill International Network for the Environment*, Glasgow: William Anderson and Sons Ltd, pp. 40–1.

Ndoye, O., Ruiz-Perez, M. and Eyebe, A., 1999, 'Non-wood Forest Products and Potential Forest Degradation in Central Africa: The Role of Research in Providing a Balance between Welfare Improvement and Forest Conservation', in Sunderland, T.C.H., Clark, L.E., and Vantomme, P., eds., *Current Research Issues and Prospects for Conservation and Development*, Rome: Food and Agricultural Organisation.

Neba, N. E., 2005, *Biological Resource Exploitation in Cameroon: From Crisis to Sustainable Management*, Bamenda: Unique Printers.

Nkwatoh, A.F., 2000, *Evaluation of Trade in Non-timber-forest Products in the Ejagham Forest Reserve of South West Cameroon*, Phd thesis submitted to the Department of Natural Resource Management, University Of Ibadan, Nigeria.

Pearce, D. W. and Turner, R.K., 1990, *Economics of Natural Resources and the Environment*, Baltimore: The John Hopkins University Press.

Peters, C.M., 1999, 'Ecological Research for Sustainable Non-wood Forest Product Exploitation: An Overview', in Sunderland, T.C.H., Clark, L.E. and Vantomme,P., eds., *Current Research Issues and Prospects for Conservation and Development*, Rome: Food and Agricultural Organisation.

Taylor, D. J., Green, N.P.O. and Stout, G.W., 1997, 3rd edition, *Biological Science*, Cambridge: Cambridge University Press.

Chapter 8

Environmental Impact Assessment
and Public Participation

Introduction

Until the mid-1990s, many development projects in African countries were implemented with limited environmental concerns. The results were catastrophic: severe environmental damage and unsustainable economic development ethics. This chapter is written within the context of the increased consensus on the part of African governments of the need to harness negative environmental impacts associated with development activities. Environmental impact assessment (EIA) must incorporate significant elements of public participation. It is recognised as a set of tools that can enhance good environmental management and governance, so as to make development sustainable. Since public participation is a slippery concept in decision making surrounding EIA, this chapter deliberates considerably on that question. Social and economic development in most developing countries currently represents a dilemma between meeting the basic human needs of an increasing population on the one hand, and conserving declining natural resources on the other. Development activities and policies concerning agriculture, dams and man-made lakes, drainage and irrigation, forestry, housing, industry, mining, power generation and transmission, waste treatment and disposal and water supply, bring about changes in the environment in which they are undertaken. These impacts can be severely adverse if the processes are not well regulated or controlled through improved project selection and more responsive planning and design. EIA is therefore introduced and discussed as a tool to chart a new course of development action, which ensures a balance between biophysical and human environments leading to the presumed state of sustainability.

Analysis of the EIA process is, however, incomplete without an articulation of public participation. Public participation in EIA is a crucial link in achieving its success. It highlights the relationship expected between the various stakeholders who have a direct or indirect interest in a development activity, the impact of which, on the environment, is the subject of examination. Ideally, public participation should be an integral component of the entire EIA process. The critical stages in which this must be undertaken fully are during the scoping phase, during the preparation of the draft EIA statement, and during the review of the draft EIA.

What is Environmental Impact Assessment?

This is probably the question with which most authors begin when addressing EIA within many contexts. Below are selected definitions of what could plausibly constitute EIA. The Southern African Institute for Environmental Assessment (SAIEA 2004:1) defines EIA as 'a process that assesses the impacts of a planned activity on the environment – physical, social and economic – providing decision-makers with an indication of the likely consequences of the development actions'. Therefore, as an integral component of the planning process, EIA enables 'potentially negative impacts to be mitigated (and positive impacts to be maximized) early in the design stages'. Through the EIA process, the developer can enhance the manner in which a project is planned, implemented and, in some instances, de-commissioned. El-Fadl and El-Fadel (2004:553) maintain that 'environmental impact assessment (EIA) was devised as a decision tool in response to grand swell of ecocentric concerns to mediate between the technocentric view of continued development and the ability to create economic growth while overcoming environmental problems'. The government of Zimbabwe defines EIA as 'an assessment of the environmental impacts of an activity, based largely on existing information and some field reconnaissance' (MoMET 1997a:5). To this end, an EIA should be undertaken during the early feasibility studies with the purpose of identifying 'likely impacts, to estimate their severity, to indicate which impacts may be significant, and to indicate what opportunities are available to avoid or minimise negative impacts and enhance potential benefits (MoMET 1997a). Hugo (2004:275) describes EIA as: 'a site specific environmental management tool designed to bring all the relevant detailed information regarding site specific development to light, which encompasses methodologies and techniques for identifying, predicting and evaluating the environmental impacts associated with project developments and actions'.

From the definitions outlined above, the following deductions can be made regarding EIA; that it is:

- a tool used to guide decision-making in ensuring that environmental as well as technical and economic considerations are taken into account
- project and site specific, thus, leading to it being highly contextual
- a process with cyclical and simultaneously linked stages
- supposed to provide monitoring, evaluation and decommissioning facets to a development project
- used to identify both the negative and positive impacts with the intention of mitigating against the negative impacts whilst enhancing the positive impacts
- applicable to both development activities and policies.

EIA is also known by other terms, amongst which: Environmental Assessment (EA), Environmental Impact Analysis (EIA), Environmental Auditing (EA), and Environmental Appraisal (EA). However, for the purposes of this publication we will stick to EIA. In addition, given our holistic definition of the concept of environment, it must be noted that modern day EIA, by default, encompasses:

- Ecological Impact Assessment (EIA) or Environmental Impact Assessment (EIA) – in their limited nature to cover mainly the biophysical components of the environment
- Health Impact Assessment (HIA)
- Social Impact Assessment (SIA).

In this regard, EIA practitioners and facilitators should always make sure that their teams include expertise from these fields. Other specialists, such as archaeologists, agronomists, environmental economists, planners, geologists and botanists, should always be consulted as per the dictates of the specific proposed development project. Their impact assessment reports will form components of the bigger EIA report.

Stages in the Environmental Impact Assessment Process

Various EIA models are presented in the literature (Biswas and Geping 1987; Hugo 2004; Weaver 2003). However, common to all these models, and to others not referred to here, is that generic stages are evident. The flow diagram in Figure 8.1 is a representation of the key stages in the EIA process. These stages share a close relationship with the generic stages in the project cycle. It should be noted that emphasis has been placed on linking the entire EIA process to public participation. Public participation in this case is taken to mean the active involvement of informed citizens including among them the disadvantaged, disempowered groups (women, children and the poor), and all other interested and affected parties in the EIA process. This presupposition is echoed in South Africa's EIA regulations that define public participation as the: 'means of furthering interested and affected parties and the public with an opportunity to comment on, or raise issues relevant to, an application for environmental authorisation, the adoption of a policy or guide in…or the compilation of an environmental management framework' (DEAT 2004:7).

Depending on the level of empowerment of the stakeholders in the EIA process through the resources available, such as levels of literacy, trust in governance issues, money, time, transportation and political power, public participation can take place as part of a spectrum that includes information, consultation and collaboration (SAIEA 2005). The views expressed above were also echoed as part of the Earth Summit major principles. Principle 10, for example, stipulates that:

Environmental issues are best handled with the participation of all concerned citizens, at the relevant level. At the national level, each individual shall have appropriate access to information concerning the environment that is held by public authorities, including information on hazardous materials and activities in their communities, and the opportunity to participate in decision-making processes. States shall facilitate and encourage public awareness and participation by making information widely available. Effective access to judicial and administrative proceedings, including redress and remedy, shall be provided (UN 1992).

In addition, Principle 20 seals the call to involve women. The principle states that 'Women have a vital role in environmental management and development. Their full participation is therefore essential to achieve sustainable development' (UN 1992).

Figure 8.1: Generic Stages in the EIA Process

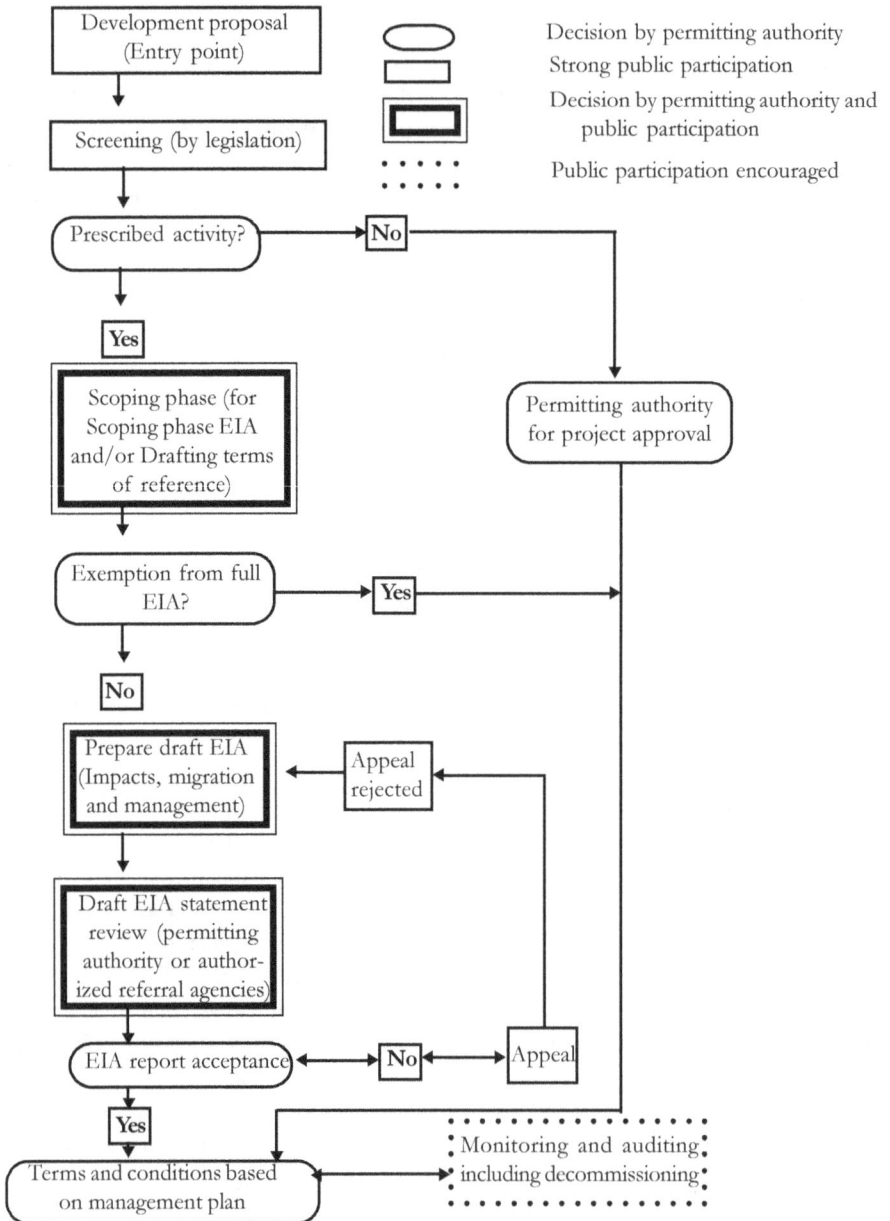

From Figure 8.1 the following key stages in EIA can be deduced: screening, scoping, preparation of the draft EIA statement (impact identification, mitigation and management plan); the draft EIA statement review as well as monitoring and auditing (including decommissioning). These and other stages are considered in turn in the next sections, including the manner in which they relate to the project cycle. The interface between the project and EIA cycles need to be carefully explained so as to add value and justification for the need to carry out EIA.

Interface Between the Project and EIA Assessment Cycles

Since there is a close link between the project and EIA cycles, it is inevitable that the constituents and the relationship that exists between these cycles be simultaneously considered. The interface in terms of the stages is summarised in Table 8.1.

Table 8.1: Interface between Project and EIA Cycles

Stage	Project Cycle	EIA Cycle
1	Pre-feasibility	Screening
2	Site selection	Scoping
3	Feasibility and feasibility report	Impact assessment and draft EIA report/statement
4	Board decision and detailed design	Draft EIA statement review and environmental management plan
5	Construction, operation and closure	Monitor, audit and decommission

Pre-feasibility (Screening) Stage

The pre-feasibility phase is the key planning stage. Certain information is known. Such information includes the basic nature of the project (for example, petroleum, irrigation or gas pipe). At times, the general site or group of alternative sites in which the development will take place (sometimes the total land area) is also known. However, at this stage, detailed designs of the proposed development are not available. The EIA activity at this project stage mainly involves screening. Usually, project type lists, which are drawn up by government, are used. Alternatives are considered and analysed. If an EIA is required, then a quick preliminary EIA can be used in consideration of alternatives. A preliminary EIA acts as an early indicator of the impacts that are likely to be significant, and helps identify of environmentally sound alternatives. A major problem with such EIA screening lists is that they do not take into account location, one of the key determinants of the nature of environmental impacts. In addition, what may appear to be a small project at the national level is not necessarily small at the local level. However, putting in place rigid screening criteria might not be the best option, as both the project and its location determine the magnitude and significance of the impact. To address this limitation, a phased screening process is perhaps the best option.

Screening can therefore be understood as a process through which a decision is reached as to whether or not to subject a project to a detailed or full EIA. This definition is, however, more applicable to situations in which legislation does not list or prescribe development projects, or in situations where border-line proposed projects (in terms of EIA requirements by law) are encountered. The screening process usually leads to one of the following decisions being arrived at, that: a detailed or full EIA is required, a limited environmental analysis (or scoping level EIA) is required, or no EIA is required. In screening, the beneficial and detrimental short and long-term effects of each alternative are compared and summarised to facilitate discussion and evaluation by interested and affected parties. Factors determining the necessity of either a scoping level or full EIA include (Hugo 2004; Murray-Hudson 1995) the nature of activity, location of the proposed development and scale (size). For example, activities such as mining, industrial, infrastructure (roads, airports, dams and power lines), agricultural activities and policies (such as resettlement, grazing, green revolution and canals) require either a scoping level, or a full EIA, to be undertaken. Activities planned in environmentally sensitive areas such as national parks, buffer zones and wetlands are subjected to similar treatment. Screening is mostly done using simple checklists for the type of an activity, and lists of environmentally sensitive areas. A checklist consists of a list of environmental parameters to be investigated for potential impacts. Checklists therefore ensure that particular environmental aspects are not overlooked during analysis. A typical screening checklist roughly estimates the likely impacts of the proposed development activity on: land, groundwater (geohydrology), surface water (hydrology), atmosphere, noise, vegetation (flora), animals and birds (fauna), human health and safety, aesthetic and cultural values as well as the socio-economic dimensions; or contains 'yes/no' questions. Two major benefits can be realised from this exercise. Firstly, it highlights potential significant environmental impacts at an early stage when alternatives can still be considered and/or when mitigation measures can be taken. Secondly, it is a cost-effective tool, which helps ensure that substantial financial and human resources are not committed to environmental analysis for development activities with few environmental impacts.

Project Site Selection (Scoping) Stage

At this stage the EIA activities centres on scoping to determine the nature of impacts associated with each possible alternative as well as the extent of the impacts, their significant, whether they are reversible or not, or whether they are direct (primary) or indirect (secondary). The public is heavily involved, and in other countries, a scoping level EIA is undertaken. The terms of reference for the finally selected project site alternative, for the full EIA, are jointly developed by the interested and affected parties or stakeholders.

Once a decision has been made – or is prescribed by law – to undertake a full EIA, the next stage is to determine its scope. The scoping exercise requires lead agencies to undertake an early and open process to determine what should be investigated, and to what extent. Scoping also helps in dealing with the type of data to be obtained, and methods and techniques to be used and the way in which the draft EIA statement results will be presented. The agencies should achieve this objective through careful consideration of existing information relevant to the assessment, as well as from the organised involvement of other agencies and consultations with the public. Issues surrounding public engagement and participation are considered further later in this chapter. The main purpose of scoping is therefore to identify the significant issues and eliminate the insignificant ones. It has been established that where scoping does not take place, delays in project implementation often occur, along with extra costs, because of time spent assessing impacts that were not identified earlier, and which eventually proved significant. Scoping therefore effectively determines the ToR of the EIA. The major issues to be considered during scoping include: definition of the activity; if not covered during screening, the identification of alternatives for development (e.g., different location or size, environmentally friendly technologies); definition of the planning time horizon; spatial scale (impacts could be *in-situ* or off-site); and coverage of human effects.

Socio-economic as well as political impacts must be incorporated into the EIA, together with their possible direct and indirect environmental impacts. On this basis, social impact assessment (SIA) is now considered an integral part of a full EIA. Impacts come in various forms, which include: significant and insignificant impacts, primary (direct) and secondary (indirect), reversible and irreversible, short term and long term, on-situ and *ex-situ* (on-site and off-site) and non-recurring and recurring (Hugo 2004; Murray-Hudson 1995). Significant impacts are outstanding impacts requiring mitigation, while insignificant impacts have negligible effects, requiring minimum or no mitigation measures. Direct impacts are a direct result of an activity, for example, the relocation of communities due to construction of a dam or a road. Indirect impacts emanate from subsequent effect caused by direct impacts, for example, the loss of business at a village shop after the community consumers had been relocated to allow for construction of a road or bridge. Generally, direct impacts are more easily identifiable than indirect impacts. The direct impacts are usually felt immediately, as compared with the indirect effects, which sometimes set in gradually.

Reversible and irreversible impacts refer to the permanent nature of impacts. Reversible impacts refer to effects that can be reversed by natural means when the project is complete, for example, re-forestation of land where soil has been borrowed and land cleared for construction of a road or dam. In this case, the borrowed areas are back-filled, and indigenous trees and grass planted, to ensure that after a period of time, they appear as natural as possible. Irreversible impacts refer to permanent effects, when the project is completed. A good example is mining, where even after rehabilitation, mine dumps remain visible and dangerous. Irreversible impacts will generally cause permanent damage to the environment.

Short-term impacts occur over a short period of time and are of a finite duration, for example, noise and dust emissions during construction. Long-term impacts occur for a very long period of time and tend to have effects well into the future, for example, reduced river flow downstream of a dam after the dam construction.

On-site impacts take place directly at the proposed development project location. Off-site impacts are those that occur far away from a project but resulting from the project. Specific impacts are therefore characterised by their extent, or by area affected. An example of on-site impact is noise vibration, and dust caused by traffic movement and blasting at a construction site. An example of an off-site impact is the shortage of a particular drug at a local clinic, due to high incidence of illness requiring the particular drug by construction personnel.

Impacts, which occur only once, are said to be non-recurring. Those which continue to occur are called recurring. An example of a non-recurring impact is the rescue of animals caught by the flooding of Kariba Dam when it filled for the first time. High incidence of seasonal mosquito bites caused by high breeding because of a dam constructed is a recurring impact. Impacts also tend to have negative, neutral or positive effects.

Positive impacts are those that bring a change for the better to a community or the environment. For example, a positive impact is the creation of jobs during construction works. Neutral impacts tend to have no impacts at all and everything continues to happen or go on as if nothing ever happened at all. Negative impacts harm, degrade or impair the ecosystem, health and quality of life of the people who live and work in the affected environment.

The techniques or methods used in achieving impact prediction and assessments in the EIA process differ quite considerably. Such methods do not provide complete answers to all questions, related to the impacts of the proposed projects. Therefore they should be selected based on appropriate evaluation and professional judgement. Higher order techniques should be selected only when those of the lower order fail to achieve the desired detail regarding a particular impact or set of impacts. Hugo (2004:278–85) identifies the following common techniques: *ad hoc* methods, checklists, matrix, composite matrix, networks, flow diagrams or models as well as map overlays and geographic information systems (GIS). Details concerning these methods can be obtained from the referenced author, see for example, El-Fadl and El-Fadel (2004), Murray-Hudson (1995).

Every proposal will bring about positive and negative change to the community in which it is located. Mitigation provides practical ways of reducing adverse impacts on the environment and social life of a community, and enhancing the benefits of the proposal. The implementation of a cement factory near a community poses health risks to the inhabitants of the community while at the same time offers employment opportunities. Mitigation measures might include the installation of efficient dust collectors, which reduce the dust being released to the atmosphere, and/or relocating settlements that are on the side of the factory where dust deposits will accumulate. Mitigation measures are better implemented at the design stage of a proposal, so that the cost of modifying the proposal will not be prohibitive.

Mitigation measures include: avoiding or eliminating adverse impacts by not taking certain actions or steps (avoidance). Limiting the degree, extent, magnitude or duration of adverse impact by reducing the size/scale of the proposal and its implementation (minimisation). Rectifying adverse impacts by repairing or enhancing the affected resources (rehabilitation or restoration which is an extreme form of rehabilitating). Compensating for loss with substitute similar resources (replacement). Typically, mitigation measures should be put in place at four generic levels in project implementation: planning, construction, operation and decommissioning stages. Mitigation measures for a water development project such as a dam during the construction phase include (MoMET 1997b):

- timing to avoid dry season discharge if possible
- timing and extent of changes in river flow adjusted to minimise disruptions, problems
- demarcation of zonation of tree clearance
- timing to minimise herbaceous plant cover
- bank stabilisation if practical
- importation of cooking/heating fuel when appropriate
- exportation of solid waste when practical
- representatives on a long-term basis in order to identify any possible remedial action
- land use planning in area of resettlement
- monitor trends in demographics, health, education, employment, crime, etc
- control of aquatic macrophytes
- health education
- reconcile recreational fishing with inshore netting and subsistence line fishing
- fisheries management plan (pelagic/inshore/recreational).

Compensation is also considered a form of mitigation, used for certain social and economic impacts, where the loss of assets, or access to resources by individuals or communities, are replaced with cash payments or alternative assets or resources. Compensation is a form of mitigation used in specific socio-economic impacts where loss of asserts or access to resources by individuals or communities may be compensated for in cash, or through the provision of alternative resources. This is a very difficult process. It is usually impossible to give full or 'fair' compensation. In many instances it is even difficult to identify those deserving compensation. The process of compensating the affected is slow, even though the impacts may be felt immediately. In assessing and evaluating compensation for lost assets or resources, the following points should be borne in mind:

- evaluation of the value of the assets or resources that will be lost from implementing the proposal

- identify the individuals or communities that should be compensated for the loss
- examine the most equitable method of compensation. The method adopted should be acceptable to all parties, and it should be such that it is self-sustaining to the beneficiaries (open-ended commitments from the proponent to the beneficiaries should be avoided)
- the cost of compensation should be incorporated into the economic analysis of the entire proposal
- the compensation plan and how it will be implemented must be outlined in the environmental management plan (EMP) which should be part of the EIA document.

It is inevitable that clear policy guidelines must be developed at either the project specific or the national level to compensate those affected and these guidelines should ideally be fair, equitable and timely. Such procedures are normally statutory. In Botswana, compensation guidelines for payment in lieu of lost land, trees, crops, structures and other fixed assets are provided by the Ministry of Local Government, Lands and Housing for use by land boards and others who require them. In Zimbabwe, the Ministry of Agriculture and Resettlement has guidelines for assessing the value of land and other assets when it desires to designate a farm for resettlement.

Project implementation usually alters the environment in one way or another. Such alterations may bring changes with certain effects. An impact often results from a change and its effects. A suitable example is the discharge of industrial effluent into a river, which reduces the amount of oxygen dissolved in water (change) resulting in fish dying (effect) affecting fishermen economically (adverse impact). Monitoring the implementation of the environment management plan becomes critical.

Findings of the full EIA exercise are presented in the form of a draft EIA report (statement). The draft should capture the following items and format: project title that identifies the type of project proposed (i.e., a multi-purpose dam, and its general location); executive summary written in a non-technical language that also includes a set of major recommendations about the way forward, including mitigation measures against negative impacts if necessary; project proponent; project description to include, but not necessarily limited to, the description of the project in terms of raw materials, processes, equipment and products, maps, flow diagrams and photographs (where applicable) and a summary of the technical, economic and environmental features essential to the project, description of existing environment to include discussion on conditions, in qualitative and quantitative terms, of the biophysical and human environments before the implementation of the project, spatial boundaries within the environment that is under consideration, and environmentally sensitive areas; project options; environmental impacts; mitigation measures; management plan (including decommissioning), key sources of data and information and list of references.

Feasibility and Feasibility Reporting (Draft EIA Statement Preparation) Stage

All EIA work should be done at this stage. It can be very expensive to do an EIA after the design stage of the project is over as the EIA may recommend a change in the whole concept or design. Post-design EIA may also lead to project cancellation after a lot of resources would have already been committed. The EIA should thus be seen as a means of prevention (anticipation) rather than cure. In as much as a project design can be forced to change due to insights drawn from the ongoing EIA study, the reverse is also true. A project design can sometimes change due to economic or engineering aspects leading to a re-orientation in the draft EIA statement preparation. It is therefore necessary for teams from both activities to be in constant and continuous liaison if the activities are to run smoothly.

Board Decision/Detailed Design (EIA Statement Review/ Management Plan) Stage

At this stage the proponent, government or donor agency makes a decision about the economic viability of the project. EIA results are considered concurrently. Approval is followed by an application for authorisation by a developer to a local or central government agency. In this way, EIA plays an important role in decision-making.

After the draft EIA statement is ready, it must undergo a detailed review process. The quality of the draft EIA statement must be such that it is of an acceptable standard (especially in the eyes of the public) and that it properly reflects the projects performance in terms of sustainability (particularly in the eyes of the permitting authority that is in many cases the government or donor). The general objectives from reviewing the draft EIA report are to: objectively evaluate the draft EIA report in relation to the ToR of the study and the quality of the findings obtained, assess the views of all stakeholders on the findings, enable decision-makers to arrive at final decisions on how to proceed with implementing the proposal, and ensure strong commitment to the implementation of the environment management plan. It is advisable that the review of the draft EIA report be done by an independent body, different from the consultants that carried out the studies and the proponent interested in achieving the implementation of the project. It is only at this point that the review can be presumed objective in assessing the quality of the data gathered, the theoretical models used for prediction of impacts, and the conclusions.

The results from the review process usually take one of these four forms: (1) complete rejection of the draft EIA – on the basis that it did not adequately cover the scope of the study or address the ToR; (2) approval of the draft EIA statement subject to major modifications – simultaneously leading to the approval signal for the development; (3) approval of the draft EIA statement subject to minor modifications and (4) approval of the draft EIA statement without any amendments – which is very rare. The last three options mean that the development project will also simultaneously be implemented as proposed. However, the developer or sub-contracted EIA experts may wish to appeal against the first two outcome decisions from the review process, as these mean that more information on impacts and mitigation for the proposed development project will be needed.

After review, the draft EIA is finalised into a manual for managing the environmental aspects of the development activity. Usually, an agreement (letter of acceptance or permit) between the development proponent and the EIA authority is signed and bound together with the final EIA statement. This agreement shows acceptance of the findings and, more importantly, enshrines the environmental management plan that outlines significant mitigation against adverse impacts. Without such a commitment, the EIA statement may simply be shelved and its recommendations ignored.

Construction, Operation and Closure (Monitor, Audit and De-commission)

Writing in the early 1990s, Kakonge (1993) identified a number of constraints on implementing EIA in Africa. Problems most noticeable then (and probably today) included inadequate environmental legislation, inappropriate institutional framework for coordinating and monitoring government activities, shortage of qualified manpower, inadequate financial resources and lack of public awareness of the need for EIAs, see for example Tarr (2003). A lot of negative impacts are likely to emerge during project implementation – its construction and operation.

Without proper monitoring and auditing, the final EIA statement may turn out to be merely a document for obtaining a permit to implement the proposal. Monitoring is required to assess whether the predicted impacts materialise, and what their severity might be. The feedback from monitoring allows for modifications in the activity and/or appropriate mitigation. Monitoring should be properly focused on (Hugo 2004; Murray-Hudson 1995): checking for the occurrence of the most important predicted environmental impacts, checking whether the mitigation measures are effective, and provide early warning about unexpected environmental impacts. Monitoring should be done at all phases of the development project. Running concurrently to monitoring is environmental auditing. An environmental audit, similar to a financial audit, assesses the performance of the development proponent. This is done in terms of the requirements specified in the final EIA statement; thus, this is specifically a compliance audit (Hugo 2004). Ideally, an independent body should do the auditing. Such an independent body could be a consultant, a representative from a regulatory body, an NGO or an informed member of the public. Both the monitoring and environmental auditing phases in EIA are linked to de-commissioning. Depending on the nature of project, de-commissioning might take three forms: ongoing, end of life or both. De-commission ensures that the environment is rehabilitated after the conclusion of operations. This is common for mining pits, landfill sites, road construction borrow pits or even dried water boreholes. The responsibility of decommissioning lies with the proponent or major beneficiaries of the project. As such, the likely costs of de-commissioning should be predicted with reasonable accuracy during the feasibility stages so that financial arrangements are made. Decommissioning resources should also be clearly indicated in the environmental management plan from the final EIA statement.

Stakeholders in the Environmental Assessment Impact Process

Stakeholders in the EIA process are of two major types: (1) those directly affected by the proposed development project (affected parties) and (2) those indirectly affected by the proposed development project (interested parties). The group of affected parties comprises the developer(s), as well as other beneficiaries in terms of aspects such as employment, improved standard of living, increased commercial activities and improved health. The other affected parties are those negatively impinged on in aspects like relocation, lost land, increased noise, pollution and traffic congestion, depending on the nature of the project. However, to identify those that will be indirectly impacted (either positively or negatively) by the proposed development project may be more difficult, and to a large extent, will be subjective. For instance, surrounding communities in which a cement manufacturing plant is located may be indirectly affected in the areas of employment opportunities and the pollutant plume from the smoke stacks. For this reason it is considered good practice to broaden the participation of persons or group of persons to include anyone who has an interest or could be marginally affected by the proposal. Other stakeholders include government (in most cases as permitting authority), environment NGOs, EIA experts and donors.It is not usually possible to involve every member of a community in a full public consultation and participation process, but it is usual to consult with representatives of the community. It is important is ensuring that those chosen for the public participation truly reflect diversity of opinion in the community. Care is required to ensure a fair and balanced representation of all views, and that the views of the poor or minority groups are not suppressed in favour of the more influential or wealthy.

EIA Legislation: Policies and Frameworks

The EIA legislative framework in Africa has improved significantly since the 1990s. At the international level, soft laws such as the Rio Declaration provide the platform for the development of EIA policies and laws. Principle 17 of the Rio Declaration states that 'Environmental impact assessment, as a national instrument, shall be undertaken for proposed activities that are likely to have a significant adverse impact on the environment and are subject to a decision of a competent national authority' (UN 1992).

At the Southern African Development Community (SADC) level, the 1996 'SADC Policy and Strategy for Environment and Sustainable Development: Toward Equity-led Growth and Sustainable Development' provides the basis for implementing Agenda 21 within the region's context (SADC 1996; SADC 2003). By 2003, all countries in SADC had either specific EIA policies and/ or framework laws in place (Tarr 2003). Table 8.2 summarises these policies and laws as well as other relevant issues concerning responsible institutions and capacity.

Table 8.2: EIA Legislation, Institutions and Capacity in Southern Africa

Country	EIA policy	Specific EIA law	Responsible institution	Capacity (6/2002)	No. of EIAs done by (6/2002)
Angola	None	Environment Framework Law, No.5, 1998	National Directorate for Environment, Ministry and Urban Planning and Environment	5 professionals	No statistics available
Botswana	National Conservation Strategy, 1990 – not strictly EIA	In progress	National Conservation Strategy	4 professionals	16 completed between 1985 and 2001
Lesotho	National Environment Policy, 1996	Environment Act, No.103, 2001	National Environment Secretariat, Ministry of Environment, Gender and Youth	3 professionals	No statistics provided
Malawi	National Environmental Policy, 1996	Environmental Management Act, No.34, 1991	Ministry of Natural Resources and Environmental Affairs	3 professionals	35 completed between 1998 and 2002
Mauritius	National Environment Policy, 1990 National Environmental Action Plan, 2000	Environmental Protection Act, No. 34, 1991	EIA Division, Ministry of Environment	7 professionals supported by environmental police	Over 800 application lodged between 1993 and 2000 but nor result on the outcomes
Mozambique	National Environmental Management Programme, 1996 – not strictly EIA	Framework Environment Law, No.20, 1997 EIA Regulations, No.76, 1998	National Directorate of EIA, Ministry of the Coordination of Environmental Affairs	8 professionals	No statistics available
Namibia	Environmental Assessment Policy, 1995	Environmental Management Bill, in progress	EIA Unit, Directorate of Environmental Affairs, Ministry of Environment and Tourism	2 professionals	82 completed between 1980 and 2001
Seychelles	None	Environmental Protection Act, No.9,1994 EIA Regulations,	EIA Unit, Ministry of Environment (located in the Office of the President)	9 professionals	No statistics available

Swaziland	Environment Action Plan, 1998 – not strictly EIA	Swaziland Environment Authority Act, No. 15, 1992 Swaziland Environmental Audit, Assessment and Review Regulations, 2000	Swaziland Environment Authority	9 professionals	An average of 2 completed each moth
Tanzania	National Environmental Policy, 1997	Environment Management Bill	National Environmental Management Council (located in the Office of the Vice-President) and local authorities	Unknown	An estimated 26 completed since 1980
Zambia	National Conservation Strategy, 1997 – not strictly EIA	Environmental Protection and Pollution Control Act, No.12, 1990 (as amended Act No.13, 1994)	EIA Directorate, Environment Council of Zambia	5 professionals	134 project briefs completed since 1997, of which 23 resulted in full EIAs
Zimbabwe	Environmental Impact Assessment Policy, 1994 National Conservation Strategy, 1987	Environmental Management Act, No., 2002	EIA Unit, Department of Natural Resources, Ministry of Environment and Tourism	9 professionals	196 completed since 1995

Source: Compiled and updated from SAIEA 2003:333–5.

Most EIA legislation in Africa prescribes projects for EIA. In Nigeria, EIA legislation, the Environmental Impact Assessment (EIA) Decree No. 86, was enacted in 1992 (Olokesusi 1998). Projects are screened using six criteria: the project magnitude, extent or scope, duration and frequency, associated risks, significance of impacts and availability of mitigation measures associated with impacts identified. In Zimbabwe (MoMET 1997a), new development proposals and substantial additions, expansions and improvements or re-construction of existing activities are prescribed as requiring EIA. The same criteria are used in Egypt and Tunisia (Ahmad and Wood 2002). In Egypt, EIA requirements are covered under Law No. 4 on Environmental Protection 1994, whilst the EIA Decree No. 362 1991 regulates EIA in Tunisia (Ahmad and Wood 2002).

Development projects requiring an EIA in Zimbabwe are prescribed based on the type of development rather than its size (Government of Zimbabwe 2002; MoMET 1994). In addition, the government can, from time to time, prescribe development activities, policies and programmes for EIA. Activities likely to affect environmentally sensitive areas such as national park estates, wetlands, *dambos* and *vleis*, productive agricultural land, national monuments and important archaeological and cultural sites are also prescribed for EIA. The full list of prescribed development activities includes, agriculture, dams and man-made lakes, drainage and irrigation, forestry, housing developments, industry, infrastructure, mining and quarrying, petroleum, power generation and transmission, tourist and recreational development, waste treatment and disposal as well as water supply.

In Egypt (Ahmad and Wood 2002), the prescribed projects are arranged into three categories: the 'black list', 'grey list' and 'white list'. Projects that require a full EIA fall under the category 'black list'. Those falling into the 'grey list' require the developer to supply considerable EIA information that will be accompanied by an environmental screening form B to enable the authority to make an informed decision as to whether a full EIA will be required or that only a limited scale (scoping level EIA) is undertaken. 'White list' projects require the developer to complete the environmental screening form A, for which only the basic project data is needed. Table 8.3 provides a comprehensive summary of the status of EIA legislation in selected North African countries.

Table 8.3: Summary of EIA Legislative Status in Selected North African Countries

Legislative parameter	Country			
	Algeria	Tunisia	Egypt	Morocco
Year enabling legislation enacted	1983	1988	1994	2003
Legal provision for EIA	Legislation & regulations	Legislation & regulations	Legislation & regulations	Legislation & regulations
Status of EIA regulations	Legislated	Legislated	Legislated	Legislated
Provisions for appeal	None	None	Legislated	None
Specification of time limits	None	Legislated	Legislated	None
Competent authority for EIA	Yes	Yes	Yes	Yes
Review body for EIA	Yes	Yes	Yes	Yes
Specification of sector responsibilities	Yes	Yes	No	Yes

Source: Modified according to El-Fadl and El-Fadel (2004:560, 562).

From the many prescriptions as to which proposed development projects require full EIA, scoping EIA or no EIA, and details of content and procedures, South Africa's 2004 EIA regulations can be considered to be the most comprehensive. Authorisation of projects for EIA is dealt with under chapter three of the EIA regulations. The chapter (made up of sections 7–21) stipulates content and procedure (DEAT 2004) for: applications; assessment of applications; screening and considerations of screening reports; scoping and consideration of scoping reports and plans of full EIA study; contents of specialist EIA reports and their procedures; contents of draft environmental management plans; consideration of draft EIA reports and issuing of environmental authorisation; and decisions of competent authority and transfer of environmental authorisations. The EIA regulations also assign lower and upper limits to certain problematic proposed development projects, both in terms of screening and/or full EIA requirements. Projects that need screening only are listed under Section 22 and fall within Category I. Those requiring full EIA are listed under Category II in Section 23.

Public Consultation and Participation in EIA

Kakonge (1998) links good governance to EIA. In his view, environmental conflicts can be resolved through the use of EIA. In this respect, EIA draws heavily on the principles of good governance: information, transparency, accountability, responsibility and participation (see for example Global Reporting Initiative 2002; IDSA 1994; IDSA 2002). Good governance presents a huge challenge to African governments, particularly accountability and transparency. As such, public consultation and engagement during EIA processes has always been limited. For example, there is no open or legislated public engagement in EIA processes in Egypt and Tunisia (El-Fadl and El-Fadel 2004). However, although varying in the levels and stages of engagement, almost all the EIA legislative framework in southern Africa makes provision for mandatory public participation and consultation in EIA (Tarr 2003). Public participation and consultation is not a direct and easy process. Readers should be warned that it is more of an art than a science. Expertise needs to be developed to facilitate public participation in EIA processes, more so, in a manner that is not viewed as encroachment on issues of governance by many politicians of the land.

Initiatives are already underway to build expertise to facilitate public participation in EIA, especially in Southern Africa. The southern African Institute for Environmental Assessment (SAIEA), a non-profit organisation with its head office in Namibia has implemented the Calabash Project to promote public participation facilitation in EIA. A pioneer group of about twenty-five experts mainly from southern Africa undertook a two-day course in public participation facilitation in EIA in Windhoek, the capital city of Namibia in May 2005. The project is funded by the World Bank. Among the participants who were from outside the SADC region were those from Kenya, Ethiopia and Cote d'Ivoire.

The SAIEA identifies four key areas that need attention in terms of public participation in EIA. These include limited capacity, political interference, participation rights and lack of experience and confidence (SAIEA 2005). Since governments are constrained in their capacity, their ability to guide and decide on EIA is compromised. On the other hand, high level political heavy weights are reported to influence decisions regarding particular EIA, especially those in which they have direct interest. Furthermore, the public, particularly those in Africa, are largely unaware of their rights in terms of EIA procedures.

This is rendered more complicated still by variances in understandings of democracy, participation and good governance across the continent, an aspect magnified by diverse religious beliefs and patterns of colonialism. Public participation processes are still an emerging feature, in which there is a lack of experience and confidence in them. After all, there are many other pressing issues that might require 'real participation' such as the need to have descent housing, HIV/Aids and poverty reduction.

Public participation is required virtually in every stage during the EIA process, or rather, at every stage when a decision has to be made about the proposed development. However, for clarity, the following are the key stages in which the public must be engaged without compromise: the scoping phase, impact identification and mitigation, and during the drafting of the EIA statement review.

Ideally, the public should also be involved during EIA monitoring and the ultimate de-commissioning of the project. However, since resources will not permit this type of engagement, capacity should be built to empower the public to carry out these activities outside the main EIA process. In fact, it is during the EIA implementation and decommissioning stages that a number of short cuts are undertaken. leading to severe negative environmental damage, as some developers will be well aware of the fact that no one will be monitoring them effectively. However, there is a need to realise the benefits and constraints of public participation as well as to ensuring adequate participation, as summarised below.

Outcome of full public participation include:

- offering all the stakeholders a sense of commitment and ownership of the proposal
- allowing for views and values which otherwise may have not been considered to be brought to bear on the proposal
- ensuring that the final proposal is the optimal one, representing the best compromise of all conflicting interests
- providing an opportunity for the public to influence project planning, design implementation, and operations in a positive manner
- offering increased public confidence in the process of decision making
- providing for better transparency and accountability in decision making
- reducing conflicts through the early detection of contentious issues.

Ensuring successful public involvement means:

- sufficient relevant information must be provided in a form that is easily understood by non-experts. Technical jargon should be avoided
- sufficient time must be allowed for stakeholders to read, discuss and consider the information and their implication
- sufficient time must be allowed to enable stakeholders to present their views
- all issues raised must be addressed and thoroughly discussed
- the selection of dates, venues, and times of meeting should be done to encourage maximum attendance
- gender integration
- good moderation is very essential.

Constraints to full public participation may be caused by:

- limitation of financial resources: Participation may be constrained by the financial situation of local people, because participation requires time away from other tasks and hence loss of income. Cash for transportation and subsistence may have to be provided for the community members that attend meetings
- the wide spread of the rural population and difficult terrains

- language and literacy level

- cultural norms: These may limit the participation of some groups, such as women in the participatory process established for the EIA process. In such cases, other options to ensure their participation have to be explored

- Heterogeneity: communities are rarely homogeneous. Although this may constitute a constraint it can, nonetheless, be exploited for the benefit of achieving the best compromise of the conflicting interest and views.

Public Participation: the Case of the Chad-Cameroon Petroleum Project

To help illustrate issues on EIA and public participation, we selected a major investment in Africa, the Chad-Cameroon Petroleum Development and Pipeline Project (Map 8.1). Given its transnational and global nature, the case study presents many challenges and insights regarding EIA and public participation in practice.

From 1993 to 1999, intensive scientific investigations and analysis, as well as public engagement and consultation with interested and affected parties, were done. The findings, amounting to twenty volumes of environmental assessment and environmental management planning documents, were incorporated into the design of

Map 8.1: Chad–Cameroon Petroleum Development and Pipeline Project

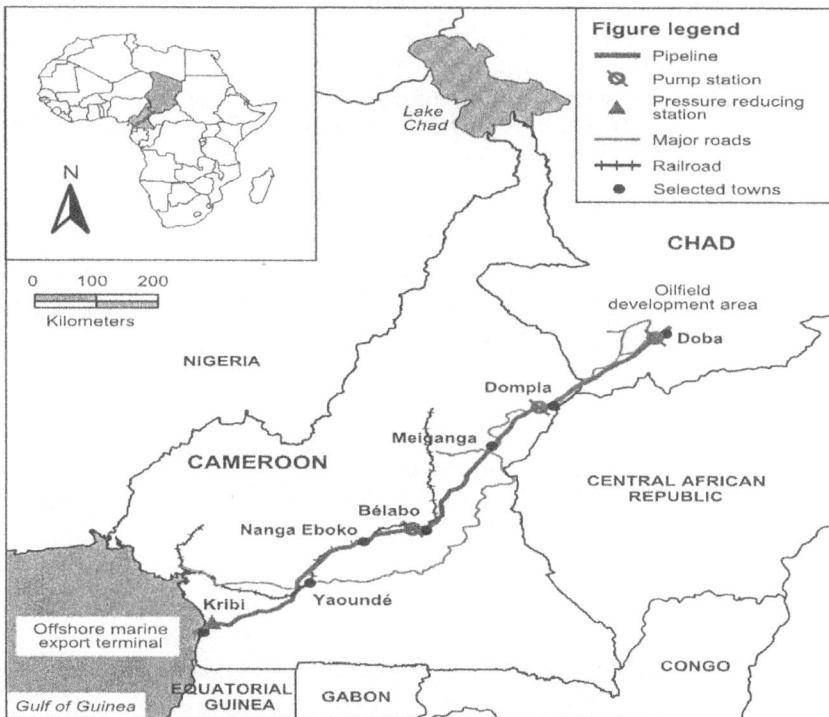

Source: Utzinger et al. 2005:69.

the oilfield and pipeline. The project was to develop the oil fields at Doba in southern Chad (at a cost of US$1.5 billion) and construct a 1,070 km pipeline to offshore oil-loading facilities on Cameroon's Atlantic coast (at a cost of US$2.2 billion). The sponsors were Exxon Mobil (the operator, with 40 per cent of the private equity), Petronas of Malaysia (with 35 per cent), and Chevron Texaco (with 25 per cent). The project was projected to result in nearly US$2 billion in revenues for Chad (averaging US$80,000,000 per year) and US$500,000,000 for Cameroon (averaging US$20,000,000 per year) over the twenty-five year production period.

The project started in 1969 when Exxon launched its programmes for oil exploration in Chad and the surrounding countries (Utzinger et al. 2005). In 1975, oil reserves were discovered in the Doba basin of Chad. Due to civil war in Chad, further explorations were halted from 1981 up until 1988. In 1993, Cameroon was brought on board as the site for offshore marine export terminal (Map 8.1). The first oil sales on the global market were recorded in December 2003. Public participation during the EIA process for the Chad-Cameroon Petroleum Development and Pipeline Project was extensive. Thousands of public comments were recorded in the facilitators' notes, videotapes and on the stakeholders' public comment database. Since 1993, about 900 village and community-level public meetings were held in the two countries. In addition to public meetings, over 145 other meetings were set up with local and international NGOs, amounting to over 250 in total. The project also made efforts to consult with organisations that had taken opposing positions to the project on a one-on-one basis. More than 700 copies of the 1997 draft EIA reports were distributed for comment to local and international NGOs, government agencies and the public. The draft EIA reports were in both English and French, the dominant languages used in the two countries. In many circumstances, the draft EIA reports had to be hand-delivered by project representatives together with a brief outlining their contents. To account for accessibility problems, several copies of the draft EIA reports were placed at seventeen public reading locations in the two countries. These were additional to customary locations that included government offices throughout the countries. Such public reading locations received over 13,000 interested and affected parties, who recorded over 9,000 comments in the notebooks provided for the purpose. Social marketing tools (Maibach, Rothschild and Novelli 2002; Nhamo 2003; Shewchuk 1994) such as making announcements through the use of local media, village level public information campaigns, and local leaders appointed as local community contacts, were utilised to raise awareness on where the draft EIA reports could be viewed for comment. The public participation programme was faced with challenges around distances covered by the project, biophysical and cultural diversity. To overcome these obstacles, the project adopted a consultation methodology built around five key principles (or guidelines). The public consultation programme had to: (1) conduct ongoing fact finding meetings, (2) take consultation to the people, (3) evolve the basis of consultation, (4) facilitate consultation with experts, and (5) comply with World Bank guidelines and directives regarding public consultation and participation. The public comments were analysed, and from the process, fifteen basic comment categories emerged. These were ranked as reflected in Table 8.4.

Table 8.4: Summary Categories for Public Comments

Category	Rank (%)
Positive views on project	22
Hiring/job opportunity/employment/training	19
Compensation/resettlement	12
Environmental impacts/pollution/leaks/spills	9
Consultation/participation	9
General project/technical/schedule	8
Project revenue/economics/ownership	6
Roads/construction/infrastructure	3
Socio-economics/cultural	3
Environmental documents	3
Safety/security/sabotage/protection of pipeline	2
Health	1
No direct relation to project	1
Project funding/Bank's role	1
Human rights/civil unrest	1
Total	100

Source: http://www.essochad.com/Chad/Files/Chad/EAESU9.pdf, accessed 1 June 2005.

Positive views indicated support for the project implementation, desire to have the project begin, and appreciation of the public consultation information provided. Issues around employment opportunities revealed that residents had unrealistically high expectations regarding the number of opportunities that the project would create. With regard to migration, the main concern raised was that there was a potential influx of migratory workers to the oilfield development area, an aspect that would result in negative socio-economic impacts to those in the vicinity. The recruitment process was also questioned, as stakeholders raised the concern that it might not be transparent, or might be manipulated along political and tribal lines.

By 2002, 12,701 people were employed; about 75 per cent of whom were nationals of Chad and Cameroon. An estimated 60 per cent of the workers from Chad and Cameroon were employed in skilled and semi-skilled jobs. Another 4 per cent were in supervisory positions. Demobilization of workers no longer needed for pipeline construction in Cameroon reduced employment of Cameroonians by over 900 at the end of the fourth quarter of 2002. While intensive construction at the oilfield facilities raised the employment figure in Chad. Wage payments to Cameroonian and Chadian workers during the fourth quarter of 2002 amounted to about US$5,700,000 and US$6,300,000 respectively.

Clarification was also sought around compensation. Although the government of Cameroon had compensation laws of 1981, these could not give a fair market value to deal with lost crops, including fruit trees. Market values had increased over time and exceeded those stipulated by the law. To this end, the project instituted a system to supplement payments to bridge the gap. In Chad, compensation was at first tackled on a one-on-one basis. Individual crops were enumerated and compensated based on government rate sheets. However, this was later modified and compensation was calculated on the basis of plot size, an element that would show equity. Community (or common property) compensation was also addressed for loss of bush land affected permanent infrastructure in particular. This came in the form of compensation in kind. Projects that resulted from this form of compensation included the construction of schools, wells, market places, roads, storage warehouses, tree planting and medical assistance.

The need to minimise resettlement was also recorded during the public consultation. As a result of an extensive pipeline routing process, no resettlement was anticipated along the route. However, a few families were to relocate fields as storage yard facilities in Cameroon. However, resettlement was inevitable in the oilfield area in Chad. An estimated 150 households were affected, and the number was significantly reduced by re-designs of the oilfield.

Emerging Issues of Concern

What emerged strongly from this chapter is that the African continent has made significant strides towards addressing sustainability principles through the application of EIA as one of the key decision making tools for approving development projects. Significant gains have been recorded in terms of establishing legislation specifically addressing EIA requirements in various countries. However, the following matters still remain slippery issues with regards to the fine-tuning of EIA procedures in the continent (El-Fadl and El-Fadel 2004; Kakonge 1993; Kakonge 1994; Kakonge 1998; SAIEA 2004).

Public Participation

The public needs to be made aware of their environmental rights. Governments need to open up when it comes to debating issues of good governance so as to encourage participation when it comes to dealing with environmental matters in the EIA process.

Local Government Blackouts

Many local authorities are not directly responsible for EIAs. Yet, most developments are implemented within their jurisdiction. In addition, local authorities have traditionally controlled development through various regional, town and country planning acts, which by their nature had considerable elements of EIA. In this regard, we recommend that efforts be made towards decentralising EIA, permitting authority to local authority assessment so that harmonisation might be worked out between town planning and EIA laws; lastly, to cut red tape.

Harmonisation of EIA Legislation at National, Sub-regional and African Union Levels

There still remain key challenges for African governments to harmonise EIA legislation at all levels. This may culminate into an African Union 'mother' EIA legislative framework. EIA laws at national levels are still highly sectoral. Yet, sub-continental frameworks (i.e., eastern, central, northern, southern and West Africa) can be put in place and eventually fed into one EIA legislative framework within NEPAD or at African Union level as an EIA Convention.

Selective Sectoral Application

Most EIAs are applied to specific development sectors, even to specific projects within the sectors perceived to have severe negative impacts (Tarr 2003). Sectors traditionally exposed to EIA in Africa include mining, petroleum and gas, as well as agriculture (but mainly limited to dams). Agricultural policies seldom receive EIA attention, yet these have the potential to harm the environment. Zimbabwe's 2000 Fast Track Land Reform Programme is one such potentially harmful agricultural programme. Fisheries and tourism projects likewise receive limited attention.

Expertise in EIA

As of June 2002, the whole of SADC had only eighty professionals managing EIA institutions (SAIEA 2004). Most tertiary institutions do not have courses in environmental management in general, and EIA specifically. Furthermore, government departments have experienced a severe 'brain drain' on a national, regional and international scale. Experienced EIA professionals often switch to better paid jobs in the private or NGO sectors. More effort is needed to encourage the establishment of courses in this arena. Resource pooling can assist in utilising the available limited EIA expertise through initiatives that seek to form coalitions between governments, NGOs, the private sector, universities and other research institutions. As is the case with the health sector, the environment and EIA must be prioritised.

Enforcement

Apart from the lack of monitoring and auditing of EIA, most legal documents do not stipulate clear monitoring and auditing procedures for EIAs and the resultant penalties to offenders thereof (Ahmad and Wood 2002). This area needs urgent attention. Regular monitoring is necessary (Tarr 2003) to ensure that developers implement the agreed-upon management plans. South Africa is one country that has taken compliance and enforcement seriously, including the establishment of environmental courts in 2004 (DEAT 2004).

Under-resourced EIA Institutions

Many in positions of authority (politics and business) still consider EIA as another unnecessary hurdle that delays development, job creation and ultimately poverty eradication in the continent. Therefore, there must be continued lobbying, particularly from peers that would realise the benefits of engaging in EIA.

Sectoral Orientation to EIA

EIAs are still undertaken and driven from a sectoral point of view. Hence, many government ministries and departments consider EIA as the sole responsibility of the ministries responsible for environment and tourism. A cross-cutting paradigm should therefore be advocated.

Logistics and Ream Management During EIA Preparation

Drawing from a large-scale EIA for the proposed Dune Mining at St. Lucia in South Africa, Weaver et al. (1996) note that EIA teams need to complement each other, not only technically, but also in purpose. The expectations and approach to the EIA should be mutually understood. All members should be mutually accountable for their joint efforts. Logistical issues, particularly around public participation, are usually seen as delaying the process.

Recognition of Potential Sub-regional EIA Promotion Initiatives

Governments should recognise and resource sub-regional EIA initiatives. One good example is the initiative by the Southern African Institute of Environmental Assessment (SAIEA), an indigenous NGO based in Namibia. SAIEA is dedicated to promoting EIA as a tool to achieve sustainable development and eradicate poverty in southern Africa. Through partnerships, SAIEA has been supporting government, development agencies, other NGOs and the private sector in the field of EIA. Some of the support mechanisms offered by SAIEA include: developing terms of reference for EIAs, independent reviewing, monitoring implementation, training (including hosting student attachments or internships), research and assisting with EIA legislation reform and formulation.

Conclusion

In this chapter, EIA and public participation processes were discussed. Various definitions of EIA and the stages in the generic EIA process were outlined. The stages covered included screening, scoping, impact identification and mitigation, draft EIA review, monitoring, auditing and de-commissioning. The interface between generic project and EIA cycles was also considered. Stakeholders in the EIA process include both the affected and interested parties. EIA legislation, particularly from southern African countries, was documented. The last part of the chapter investigated public participation with a case reference from the Chad-Cameroon Petroleum Project. In sum, the chapter presents a toolkit for undertaking EIA and public participation.

Revision Questions

1. What is EIA?
2. What are the generic stages in the EIA process?
3. How are the project and EIA cycles related? How does this help to understand the EIA process better?
4. From the text, what is public participation?

Critical Thinking Questions

1. How would you address issues of corruption with regard to the approval of EIAs if you were to be a senior official in one of the government offices responsible for this task?
2. Identify a proposed development project that has failed to take off due to EIA require-ments. What EIA issues have been raised by the regulating authority, which led to the delays? How best could you have addressed the concerns stopping the implementation of the development project?
3. Identify and discuss key legislative provisions for EIA and public participation in your country. What challenges exist?

References

Ahmad, B. and C. Wood, 2002, 'A Comparative Evaluation of the EIA Systems in Egypt, Turkey and Tunisia', *Environmental Impact Assessment Review*, Vol. 22, pp. 213–34.

Biswas, A.K. and Geping, Q., 1987, *Environmental Impact Assessment for Developing Countries*, London: Tycooly International.

DEAT, 2004, *Environmental Impact Assessment Regulations*, Pretoria: Government Printer.

El-Fadl, K. and El-Fadel, M., 2004, 'Comparative Assessment of EIA Systems in MEAN Couintries: Challenges and Prospects', *Environmental Impact Assessment Review*, Vol. 24, pp. 553–93.

Global Reporting Initiative, 2002, *Sustainability Reporting Guidelines*, Boston: Global Report-ing Initiative.

Government of Zimbabwe, 2002, *Environmental Management Act*, Harare: Government Printer.

Hugo, M.L., 2004, *Environmental Management: An ecological Guide to Sustainable Living in South-ern Africa*, Brooklyn Square: Ecoplan.

IDSA, 1994, *Corporate Governance King I Report – 1994*, Parklands: Institute of Directors in South Africa.

IDSA, 2002, *Corporate Governance King II Report – 2002*, Parklands: Institute of Directors in South Africa.

Kakonge, J.O., 1993, 'Constraints on Implementing Environmental Impact Assessments in Africa', *Environmental Impact Assessment Review*, Vol. 13, pp. 299–308.

Kakonge, J.O., 1994, 'Monitoring of Environmental Impact Assessments in Africa', *Envi-ronmental Impact Assessment Review*, Vol. 14, pp. 295–304.

Kakonge, J.O., 1998, 'EIA and Good Governance: Issues and Lessons from Africa', *Environmental Impact Assessment Review*, Vol. 18, pp. 289–305.

Maibach, E., Rothschild, M.L. and Novelli, W.D., 2002, 'Social Marketing', in K. Glanz, B. K. Rimer, and Lewis, F.M., eds., 3rd edition, *Health Behavior and Health Education: Theory, Research and Practice*, San Francisco: Jossey-Bass, pp. 437–61.

MoMET, 1994, *Environmental Impact Assessment Policy*, Harare: Government Printers.

MoMET, 1997a, *Environmental Impact Assessment Guidelines*, Harare: Government Printers.

MoMET, 1997b, *Environmental Impact Assessment Training Manual*, Harare: Government Printers.

Murray-Hudson, M., 1995, *Introduction to Environmental Impact Assessment – A Course Book*, Gaborone: Department of Environmental Science, University of Botswana.

Nhamo, G. 2003, 'Social Marketing: Can it Enhance Re-use and Recycling at Household Level in the SADC region?', Windhoek: Environmental Education Association of Southern Africa.

Olokesusi, F., 1998, 'Legal and Institutional Framework for Environmental Impact Assessment in Nigeria: An Initial Assessment', *Environmental Impact Assessment Review*, Vol. 18, pp. 159–74.

SADC, 1996, *SADC Policy and Strategy for Environment and Sustainable Development: Toward Equity-led Growth and Sustainable Development*, Gaborone: Southern African Development Community.

SADC, 2003, *Draft SADC Regional Indicative Strategic Development Plan (RISDP)*, Gaborone: Southern African Development Community.

SAIEA, 2004, *Environmental Impact Assessment in Southern Africa: Summary Report*, Windhoek: Southern African Institute for Environmental Assessment.

SAIEA, 2005, *A Pilot Training Course on Public Participation in Environmental Impact Assessment in Southern Africa*, Southern African Institute for Environmental Assessment, Windhoek.

Shewchuk, J., 1994, *Social Marketing for Organisations Factsheet*, Ontario: Queens Printer.

Tarr, P., 2003, 'EIA in Southern Africa: Summary and Future Focus', in P. Tarr, ed., *Environmental Impact Assessment in Southern Africa*, Southern African Institute for Environmental Assessment, Windhoek, pp. 329–37.

UN, 1992, *Report of the United Nations Conference on Environment and Development*, New York: United Nations Secretariat.

Utzinger, J., Wyss, K., Moto, D.D., et al., 2005, 'Assessing Health Impacts of the Chad-Cameroon Petroleum Development and Pipeline Project: Challenges and a Way Forward', *Environmental Impact Assessment Review*, Vol. 25, pp. 63–93.

Weaver, A., 2003, 'EIA and Sustainable Development: Key Concepts and Tools', in P. Tarr, ed., *Environmental Impact Assessment in Southern Africa*, Windhoek: Southern African Institute for Environmental Assessment, pp. 3–10.

Weaver, A.V.B., Greling, T., Van Wilgen, B.W., et al., 1996, 'Logistics and Team Management of a Large Environmental Impact Assessment: Proposed Dune Mining at St. Lucia, South Africa', *Environmental Impact Assessment Review*, Vol. 16, pp. 103–113.

Chapter 9

Understanding Environmental Education

Introduction

Environmental problems have been subjects of popular discussions and debates from time immemorial. The result is a long history of educational campaigns and movements that eventually forged the way for the inclusion of environmental concerns into national and international policy and decision making. Environmental education has been recognised as a tool or mechanism for providing lasting solutions to these problems. This chapter starts with an exploration of some of the campaigns and movements. It then focuses on the development, forms, definitions, goals, objectives approaches and programmes of environmental education.

Ancient Landscapes Towards Modern Environmentalism

Swan (1975) and Cunningham et al. (2003) present a comprehensive report on the growth and development of environmentalism in the US. Swan suspected that even as far back as when man started using metals, dire predictions were being made about the survival of humankind. The ecologists of the day were already warning of the perils of tampering too much with the natural scheme of things.

Colonial times witnessed the birth, in the US, of the conservation movements. During that period, a few laws were passed to prohibit the burning of forests and to protect various wildlife species. Thomas Jefferson and George Washington warned against destructive land practices, as they feared an adverse economic impact. Later, Alexander Wilson and John James Audubon expressed concern about the extinction of certain wildlife species and sought their protection. Ralph Waldo Emerson and Henry David Thoreau were interested in the relationship between man and nature, and what could happen should there be dissonance. But despite the efforts of these and others, conservation still meant nothing to many people in the US, as they still believed that natural resources were inexhaustible. It was not until the publication in 1864 of George Perkins Marsh's book *Man and Nature* that this myth of resource inexhaustibility came under careful examination.

By the 1930s, several efforts had been made to conserve wildlife and wilderness by establishing hunting regulations and designating parks and refuges in the US. Yellowstone National Park was created in 1872, as the world's first national park. It is regrettable that up until today, most of the rural populations of developing countries, and even some members of the elite community, still believe in the inexhaustibility of natural resources. But before the 1930s, in the US, John Muir, John Burroughs and John Wesley Powell had contributed enormously to the concept of resource conservation and to the cause of aesthetic and wilderness conservation. Muir was a naturalist who was instrumental to the founding of the Sierra Club. Burroughs, a famous outdoor writer, popularised the study of nature. Powell is best known for his report to Congress on the management of the arid west, and his exploration of that region.

Educational efforts, following the conservation movements, gave rise to the educational movements. Aldo Leopold, a forester who later became an ecologist, was famous in this domain. He added an ethical dimension to the concept of wilderness preservation conceived by Muir and the others, and did much to make game and wildlife management a recognised profession.

Prior to Leopold's ethical concept, the justification of most conservation efforts was based on aesthetic, religious or economic grounds. Leopold pointed out that humans share the earth with other life forms; and that the well-being of each life form hinges on the well-being of every other life form. He based the land ethic on the ecological premise that since humans share the environment with all other life forms, it is their responsibility to maintain it in the best interests of the total community of life.

Of the many educational movements that emerged in Leopold's time, three were prominent: nature education (nature study), conservation education, and outdoor education. These movements, or subjects, are related and partially overlap. They are considered to be the forerunners of environmental education because of their contribution in content and methods to the subject. By 1971, many institutions of higher learning in the US were already offering environmental education courses.

In the UK, the earliest focus on environmental education was in 1943, when the Council for the Promotion of Field Studies was formed. This later became known as the Field Studies Council (Boulton and Knight 1996). The Field Studies Council heralded an upsurge of environmental activities related to the countryside (Wheeler 1981).

The need for environmental education grew more urgent in the 1960s as a result of the increasing evidence of environmental degradation. In 1960, the World Wide Fund (now known as the World Wide Fund for Nature, or the WWF) was launched to raise the level of public concerns about the rapid disappearance of species (Boulton and Knight 1996). Soon after, in 1963, Rachel Carson's book, *Silent Spring*, was published. This book drew popular attention to the potential dangers of the excessive use of pesticides. A few years later, a delegation from Sweden raised some of these concerns during a UN meeting. This initiative, together with the urgent American interest in the environment, which found expression in the United States National Environmental Policy Act, with its early phraseology about inter-generational equity, which came into effect in 1970 (Wood 1999), influenced the conference on human environment.

The Internationalisation of Environmental Education

The term environmental education was first recorded following a conference held under the auspices of the International Union for the Conservation of Nature and Natural Resources (now the World Conservation Union or IUCN) in Nevada in 1970. It first appeared on the world agenda after the United Nations Conference on the Human Environment held in Stockholm in 1972. The conference designated the 5 June as World Environment Day, and encouraged governments to celebrate that day every year by organising activities aimed at raising environmental awareness or ensuring environmental protection (Vinke 1993). Representatives at the conference recommended that the UN establish an international environmental education programme. In response to the recommendation, UNESCO sponsored a series of environmental education workshops and conferences around the world. Representatives from member countries met in Belgrade, in the former Yugoslavia in 1975, to outline the basic definition and goals of environmental education.

Another international conference was held in 1977 in Tbilisi in the former Soviet Republic of Georgia. More than sixty nations were represented at that conference. Delegates from the two conferences ratified the definition and objectives of environmental education. The conference recommended that environmental education should be directed to the general public at every age, at all levels of formal education, at specific occupational and social groups (administrators and planners, industrialists, agriculturists and so on) through formal and informal education. This is in consort with the popular Brundtland Report (Box 9.1) which identifies both formal and informal environmental education as tools for sustainable development. The report enumerates some informal methods that could be employed to promote environmental education; these include special interest groups and on-the-job training. It also stresses the need for a multidisciplinary approach to the subject.

Box 9.1: The Bruntland Report

In 1984, the United Nations established the World Commission on Environment and Development, otherwise known as the Brundtland Commission. The mandate given to the commission recognised the need to link long-term environmental strategies to those of development. It urged the commission to recommend ways by which concern for the environment could result in greater cooperation among developing countries and between countries at different stages of economic and social development. In 1987, the commission published its report, *Our Common Future*, also known as the Brundtland Report.

Defining Environmental Education

The definition of the environment was discussed in Chapter One. Education could be defined simply as a process that aims to effect a positive change in behaviour. It could also be defined as the behavioural change due to exposure to some kind of systematic training or experience. It is considered to be education for sustainability (IUCN 1993). It is a process of empowerment through equipping people with information in a manner that permits them to gradually, systematically and purposefully transform information into useful bodies of knowledge. It further equips them with the appropriate skills that enable them to apply the knowledge to the transformation of resources into goods and services for the betterment of society. Note the use of the expressions 'positive change in behaviour' and 'betterment of society'. These are used to distinguish good education, because a negative connotation would mean using the knowledge and skills acquired to instead produce intentionally what could be used to destroy society.

Environmental education was defined by delegates of the Belgrade and Tbilisi conferences as 'a process aimed at developing a world population that is aware of and concerned about the total environment and its associated problems, and which has the knowledge, attitudes, skills, motivation and commitment to work individually and collectively toward the solution of current problems and prevention of new ones'. Put simply, environmental education is a process aiming to improve the quality of life by empowering people with the tools and mechanisms they need to solve and prevent environmental problems. It could also be defined as a subject that aims at changing the attitudes and perceptions of individuals with regard to the treatment and use of natural resources. It aims to inculcate in the individual a sense of accountability in the use of the limited natural resources.

Environmental education is concerned with the creation of conditions that favour modifications of attitudes or behaviour that have negative impacts on the environment. Environmental education increases awareness about issues by interrogating attitudes, perceptions, values and beliefs. It helps individuals to evaluate and clarify their feelings about the environment, and how they contribute to environmental problems. It also helps them to understand that there are conflicting values among people and that these conflicts must be addressed if environmental problems are ultimately going to be prevented or solved. In the same mode, WWF (1988a) states that it enables people to understand, analyse and evaluate the relationship between people and the environment.

Touré (1993) asserts that environmental education is not only an instrument in the fight against environmental problems, but also a precondition for the rational management of environmental resources needed for human survival. He further postulates that the efficacy of environmental education depends on the interest of the topic chosen, and on the role it can play in improving the income, hence the socio-economic situation, of the populations concerned. It is the demonstration of the link between economic development and the protection of environmental resources that very often convinces the local population to take action. Communities will support environmental programmes only if they reflect their local beliefs, values and ideologies (World Bank 1992).

Scope and Forms of Environmental Education

Environmental education has hardly been given the scope and dimension it deserves in any single situation, due largely to disputes that exist about its goal. Schneider (1993) attributes this to the fact that in any given local or national situation, environmental concerns of individuals, groups of people, and even institutions, tend to be limited rather than embracing. Also, the capacity of these actors is limited by their place in society and by the means at their disposal. Hence, there is always a tendency in site-specific situations, for example, in a conservation project setting, to limit environmental education to areas such as agro-forestry, conservation and hygiene.

According to Dunlop (1993), environmental education covers forms of understanding and dimensions of behaviour. It must address values and beliefs regarding the political economy and quality of life in a society. It is also concerned with the distribution of power at all levels, not only within nation states, but across national boundaries, from supranational organisations or transnational companies. It must also concern itself with the fine details of the natural world, understanding the importance of maintaining intact all the links in the chain of being and survival.

Martin (1990) summarises these thoughts in one of his guiding principles of environmental education: environmental education should consider the environment in its totality – natural and man-made, ecological, political, economic, technological, social, legislative, cultural and aesthetic. This means that people should be able to develop an understanding of the ecological processes that govern life on earth; an understanding of the geo-morphic and climatic patterns that influence living things and human activities; an appreciation of the social, economic and cultural influences that determine human values, perceptions and behaviour; and an awareness of an individual's own personal relationship with the environment, as a consumer, producer and sentient member of society (WWF 1988).

There are three forms of environmental education: education about the environment, education from or through the environment, and education for the environment. These form what is sometimes referred to as the trinity of environmental education (Figure 9.1), which suggests inseparability. It also points to holism: looking at the 'whole' rather than at separate parts. The holistic approach is advocated by many experts. The reasoning is that none of the three forms can stand out separately, and significantly merit the name of environmental education. All three must be blended into a single 'whole' to give depth and breadth to the subject. Let us, for the purpose of study, consider them separately.

Education about the Environment

Here the main consideration is the acquisition of knowledge about the environment, including the various components, and how they interact and interrelate. Attention is paid to the problems facing the environment, their causes and possible solutions. This form brings us to stage one, which can be called environmental science or environmental studies, an important part of environmental education.

Figure 9.1: Trinity of Environmental Education

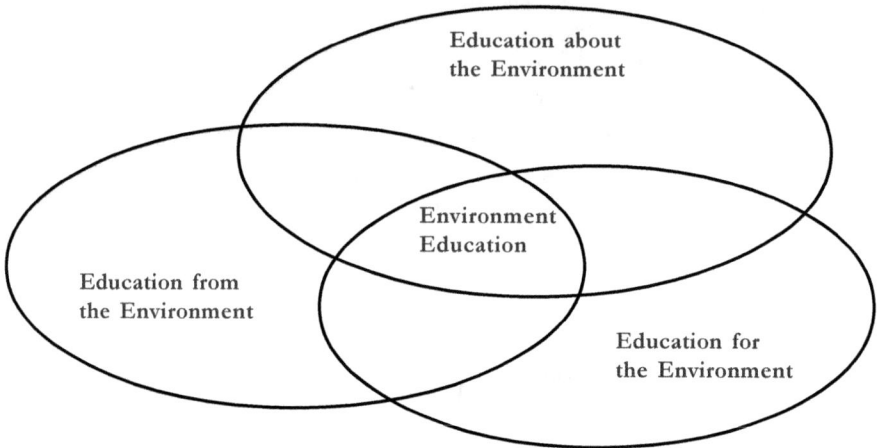

Education from the Environment

Put simply, this means using the environment as a teaching resource. Students carry out observations or experiments on different environmental aspects. Over and above gaining knowledge, appropriate skills and concepts are developed in the course of conducting experiments. These skills and concepts help equip and prepare the individual for environmental commitment. This is clearly an extension of the first form, and brings us to stage two.

Education for the Environment

This form brings us to stage three, the last stage. It could be described as the stage of environmental commitment and action. This is where you can safely hope to achieve the goal of environmental education. Individuals use their knowledge and skills acquired to improve the quality of the environment in one way or the other. This should, however, happen voluntarily, without pressure from anyone or any external factor such as the threat of law-enforcement or promise of a reward. Should these be the motivating factors behind any action taken by an individual, we cannot safely say that the individual has acted because he or she has become environmentally educated.

Objectives of Environmental Education

The goal of environmental education is not to confer new behaviour on to individuals, nor to brainwash them into thinking in a particular way. Rather, it is to help them to learn how to think, including how to investigate and analyse, how to solve problems, make decisions, weigh opinions, and align values with personal actions.

Environmental education stresses five objectives: awareness, knowledge, attitudes, skills and action, which can be remembered easily by the acronym AKASA (Figure 9.2). These five objectives are synchronised with the three levels of learning: cognitive (knowledge), affective (attitudes and feelings), and psychomotor (physical or motor skills).

Figure 9.2: Interactions of the Objectives of Environmental Education

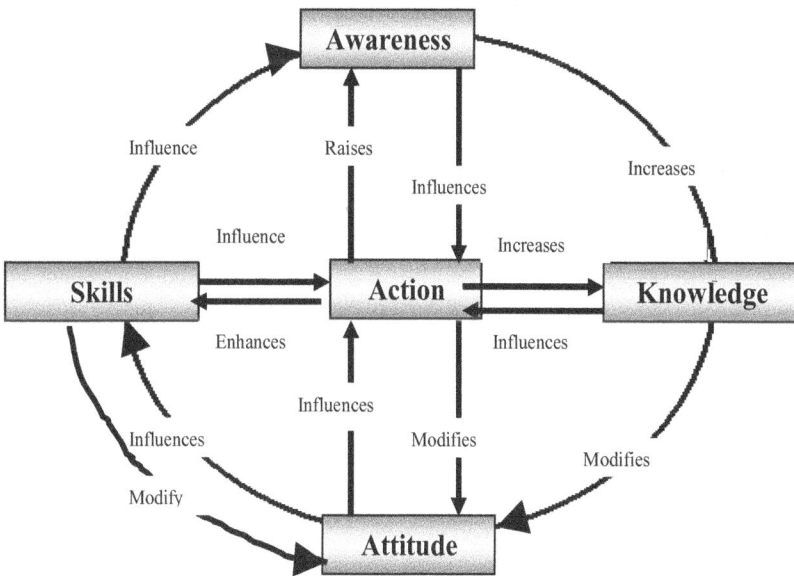

These objectives interact with one another to finally produce the core effects in the 'action' (Figure 9.3). It can be deduced that the intensity and impact of an action increases or decreases with the frequency of the action. This is largely influenced by the factors (including external influences), and level and direction of interactions that lead to a given action in a given situation.

Awareness

To help people become aware of the problems affecting the environment, their causes and possible solutions. The task here is to bring the people to genuinely examine their relationships with their environment with a view to identifying and appreciating the impact of such relationships on the environment.

Knowledge

To enable people to gain a better understanding of the various components of the environment, how they interact, interrelate and inter-depend with one another, and the natural and human factors that can cause a dissonance or total disruption of the interconnectedness and balance of the components. But the acquisition of knowledge is not in, or of, itself sufficient; the knowledge should be able to lead on to the formation of appropriate environmental attitudes and behaviour.

Attitude

To prepare and guide people to acquire environmentally appropriate social values, strong concern for, and sensitivity towards, the environment, and the motivation to actively participate in its conservation. People's values are the products of their perceptions, and the determinants of their actions. Their attitudes towards the environment are a combination of how they perceive it and how they think they ought to treat it. Noibi (1990) conducted research with a sample of fifty-five teachers, drawn from a population of 150 teachers who had completed a course in environmental education at the Lagos State College of Education in Nigeria. The results showed a significant correlation between environmental knowledge and environmental attitude ($r = 0.20$, < 0.05), indicating that knowledge about the environment could lead to the formation of appropriate environmental attitudes.

Skills

To equip people with the basic skills necessary to analyse and solve or prevent environmental problems, skills are the tools that prepare people for action, which can be triggered by an appropriate level of motivation generated from positive environmental attitudes.

Action

To enable people to develop a sense of responsibility and accountability towards the environment, and a commitment to take appropriate actions or participate actively to address currently identified environmental problems. As shown in Figure 1.2, action (also known as participation) occupies a central position. It is often observed following environmental education interventions, as proof of the effectiveness of such initiatives. Noibi's (1990) research showed a significant correlation between intended environmental action and actual environmental action ($r = 0.02$, <0.05), pointing to the possibility of intentions, leading to desired goals. But environmental attitude was shown to have no correlation with actual environmental action, whereas its correlation with intended environmental action was highly significant ($r = 0.57$, $<.01$). A simple explanation for this could be that external factors might have had, and do have, an influence in deciding whether or not the actual environmental action is realised.

Environmental Awareness and Sensitisation

A majority of environmental agencies and organisations set the goal of raising awareness or sensitising target populations to particular or general environmental concerns and issues. Although these are two different sub-targets in terms of level, with sensitisation occupying a higher echelon (Figure 9.3), the agencies and organisations invariably aim at the same ultimate target: action. However, it should be noted that it takes greater effort and more expertise to sensitise than to raise awareness.

Raising the awareness of a target population to environmental issues can lead to, but is no guarantee of, their sensitisation. To raise environmental awareness is to make the target population conscious of the prevailing environmental issues and possible solutions; whereas to sensitise them is to stimulate their feelings in such a way that they develop concern and a responsible attitudes for the environment. Sensitisation could also be defined as a process by which people are enabled to develop a set of beliefs, values and attitudes that can positively influence their perception, treatment and use of the environment and its resources.

Figure 9.3: Awareness and Sensitisation Occupy Different Levels

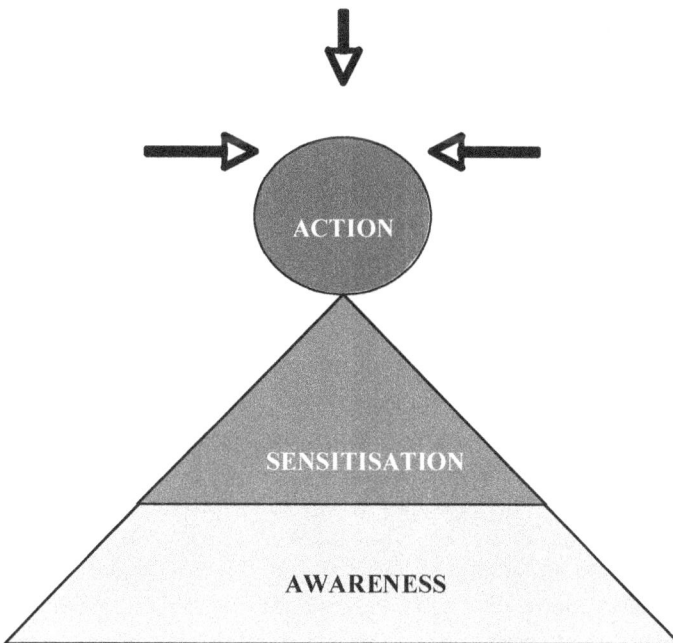

Sensitisation can lead to the cultivation of positive environmental attitudes, which, in normal circumstances, translate into positive environmental behaviour, best expressed in actions. To reach this final stage is not easy; it depends on the level of sensitisation of the individual, which increases with age, not merely in biological terms, but with respect to an accumulated attachment to particular ideas and values (Patterson 1993). It may occur as a result of subtle but powerful influences, such as advice, and examples of the family and peer groups, rigorous norms and belief systems of the community, patterned indoctrination through programmes of the mass media, schools and similar institutions, or emerging economic and related opportunities. This suggests that formulating appropriate strategies for changing or modifying the general attitude of a particular group of people towards the environment may require a historical journey in time and space to establish what the attitude was originally, and why, how it evolved over time into its present form, and the factors responsible for its evolution. Knowledge of all this is crucial to the formulation of an appropriate sensitisation strategy. A multiplicity of methods could be employed to bring about sensitisation and the desired change in attitude.

The effort required to sensitise an individual increases with the age of that individual. This explains why experts advise that environmental education be given to people in their childhood, while they are still in elementary school, so that by the time they reach the age where they tend not to listen to advice, their values and attitudes will have been cultivated already, and will remain an important part of their lives. On the other hand, starting environmental education with people at adulthood presents great difficulties, especially if the values and attitudes they have developed from childhood are in opposition to the tenets of the new discipline. These childhood experiences that have become oppositional forces, are enhanced by the cultural, economic and political climate in which people are born and nurtured.

Organisational and Institutional Arrangements for Environmental Assessment

Although many countries have not yet incorporated environmental education into their school systems, there are many institutions and organisations that provide excellent examples of environmental education programmes, aimed at both schools and the general public.

Botanical and zoological gardens offer opportunities for broad-based environmental education (Boulton and Knight 1996). Both children and adults benefit from guided tours, outreach materials and other educational services provided by these establishments. Museums, too, have changed their traditional focus on exhibits to include special educational programmes.

National parks and other protected areas are now playing an increasingly significant role in the promotion of environmental education. Many have information centres, which carry out educational activities ranging from audio-visual shows, interactive displays, printed materials to educate the casual visitor tourist, and display notices, whilst special programmes are arranged for school groups. Some examples of the above establishments also provide opportunities for teachers to receive training in environmental education. The Bronx Zoo in New York offers a good example of well designed programmes of teacher training in this domain. A number of institutions offer specialised courses in environmental education tailored to the needs of the participants, for example the International Centre for Conservation in the

UK. Some universities in both developed and developing countries are now offering certificate, diploma and degree courses in environmental education.

Prerequisites for the Implementation of Environmental Education

Before implementing an environmental education programme, it is absolutely necessary to carry out certain tasks, which could help place the implementer in a position to make informed decisions. These tasks consist of carrying out elaborate studies at the early stages of the programme. Wood and Wood (1985) elaborate the stages involved in the design of conservation education programmes, applicable to environmental education programmes, detailed below.

Assessment of the Environmental Situation

The first task is to carry out an assessment or appraisal of the environmental situation. This means identifying and assessing the various components of the biophysical environment and the various human activities that are likely to have an impact on them, positively or negatively. By extension, it involves an identification of the prevalent environmental problems, the causes of the problems, and the potential solutions.

Identification of the Target Audiences

A target audience can be described as a group of individuals who, directly or indirectly, are affected by or contribute to the identified problems, or who could be instrumental in the implementation of possible solutions. These could be men, women and youth; farmers, hunters, fishers and harvesters of forest products, or community institutions such as the Village Traditional Council, Women's Group and Youth Association.

It is necessary to understand the role and level of influence of each of the target audiences in the community, how they interact with each other, and their perception and level of awareness of environmental problems. Assessing the level of education within each target audience will also be useful in determining how best to deal with its members. It is equally necessary to understand the community's methods of communication, including both language and local idioms, and how each target audience reacts to new ideas or innovations.

Formulation of a Working Strategy

Most programmes fail. This is due largely to unrealistic and ill-defined working strategies, which help illuminate the way towards the right decisions, the right directions and the right paths.

Box 9.2: Framework for Developing an Environmental Education

Awareness/Knowledge

- programme and/or activity goals and objectives clearly defined and stated
- benefits of activity to communities clearly demonstrated and/or articulated
- environmental, health and other relevant issues adequately covered
- (these should include both general and specific ones).

Skills

- good communication skills promoted amongst staff and with and among communities
- simple analytical skills developed in community members (e.g. to analyse environmental problems and issues also in a more scientific way)
- practical activities which are culturally acceptable, intellectually, physically and socio-economically rewarding to communities and gender sensitive.

Implementation Tools

- communication tools for awareness raising and sensitisation
- cost-effective, attractive and culturally acceptable mass media tools, which are
- intellectually and socially meaningful, rewarding to communities and gender sensitive
- identification of key target groups
- identification of potential collaborators and areas of collaboration
- inter-personal communication tools that are culturally acceptable intellectually and socially meaningful and rewarding to communities and gender sensitive.

Mechanisms

- mechanism(s) for ensuring regular activities in communities
- activity sustainability mechanisms that include both short and long term dimensions
- mechanism(s) for targeting various categories of communities (e.g. receptive/non-receptive ones, those with schools and those without, those with roads and those without)
- mechanism(s) for exploiting/minimising prevailing/emerging opportunities/threats which could mean encouraging internal and external collaboration, avoiding duplication of efforts, guaranteeing the removal of potential obstacles to activity implementation, acquiring additional information, knowledge and skills, and securing additional material and/or financial support)
- mechanism(s) for monitoring and evaluation (e.g. regular feedback from selected community groups).

Source: Inyang 2007.

A strategy could be defined as a concrete plan of 'what', 'how', 'with whom', and 'when'. Action should be geared towards the intended goal. It clearly defines the goal and objectives of the programme, the target audiences and appropriate methods to be employed. It therefore means that before formulating the working strategy, the goal, objectives and methods need to be properly defined. A proper definition should not be formulated only in terms of wording, but also by linking between the categories. It should be clear how the methods, together with the activities on which they are built, contribute to the achievement of the objectives, and how the objectives, in turn, lead to the goals.

Other important elements are those that guarantee the sustainability of the programme. It has been generally observed that many environmental education activities fail due to a lack of mechanisms, also of sustainability. In a consultancy assignment for WWF-Coastal Forests Programme in Cameroon, Inyang (2007) developed a generic framework for the formulation of an effective environment education strategy. An important feature of the framework is a number of mechanisms for programme sustainability (Box 8.2). This framework was used in workshop settings to formulate environmental education strategies for four different projects, with obvious positive results.

Formulation of Environmental Messages

This is a very important, but perhaps the most difficult, task of the environmental education. Each target audience needs specific environmental messages communicated in particular ways, using appropriate media in particular situations in order to make the programme meaningful and effective.

Some messages may be used to address the entire village community or the Village Traditional Council, while others could be aimed at specific target audiences, like hunters, fishers, farmers and harvesters of non-timber forest products. Addressing a well-tailored environmental message to a wrong target audience may either render it ineffective, which is simply a wasted effort, or generate a bitter reaction from the audience. For example, giving a sensitive message on hunting intended for community decision-makers directly to a group of hunters may provoke their antagonism.

Ngome (1992) points out that the most fundamental question concern the kinds of things people should be told to make them realise that there is a serious problem, and that they will have to contribute to a solution. The core message to bring about awareness will be somewhat different from the message that is meant to bring about action. Even when people finally realise that there is a problem, it is usually difficult for them to accept responsibility for that problem. To avoid negative reactions, it may be practical to down-play their contribution to the problem and highlight the importance of their contribution to a solution.

By the same token, in formulating an environmental message, as much as possible, the positive should always be highlighted. This may seem difficult at first, as the easiest way, in most cases, is to employ the word 'don't'. But it can be effective to minimise using the 'don't' word. The target audience should be left to make its own decision about whether or not to continue with an environmentally unfriendly activity. In other words, the word 'don't' should be allowed to come from its members. There is increasingly common experience that once members of a target audience feel that a decision has been imposed on them, they will not implement it. Instead, it may raise suspicion, unless they are sufficiently convinced that implementing it is in their own best interests.

Communication of Environmental Messages

Communication is an extremely important process in environmental education, especially if the effort is to ultimately influence a desired change in behaviour. Adzeyuf [undated] defines communication simply as the process by which one or more people are engaged in the exchange of ideas, knowledge, facts, feelings and impressions in such a way that the receiver can gain a clear understanding of the message and its intention. Seiler and Beall (2002) define communication as the simultaneous sharing and creating of meaning through human symbolic action. The latter definition stresses the importance of meaning: for what is a message without a shared meaning?

Messages are sent and received through channels. Horn and Rogers (1975) define a communication channel as the means by which a message gets from the source to the receiver. The most effective strategy to inculcate environmental education into the social system – school or local community – is to combine mass and interpersonal communication channels, as their combined effect is greater than that of either used separately (Chaffee 1972; Rogers and Shoemaker 1971; Rogers 1973). While mass media – radio, television, newspapers – are the most effective in increasing levels of awareness and general knowledge about environmental issues, interpersonal channels are most effective in convincing people to change what they actually do in practice (Horn and Rogers 1975). The purpose of communication, therefore, is to bring about social change, which Rogers and Shoemaker (1971) define as a process by which alteration occurs in the structure and functioning of a social system. They suggest three sequential stages in the change process: invention: creation or development of new ideas; diffusion: communication of new ideas to members of a given social system; and consequences: changes that occur as a result of the adoption or rejection of the innovation.

In any social system, the members are linked together and influence one another to produce a diffuse effect. A diffuse effect is like a chain reaction. It is the cumulative increase in the degree of influence upon an individual to adopt or reject an innovation, as a result of the increasing rate of knowledge and adoption, or rejection, of the innovation in the system. Once the adoption rate starts to increase, the social climate encourages others to 'get on the bandwagon'. It is important to note that the pace at which an innovation is diffused is a function of its complexity, conspicuousness, compatibility with existing practice, and the existence of appropriate media of communication (Ritson 1977). The more complex, less conspicuous or less compatible with existing practice an innovation is, the slower its diffusion rate, and the other way round. Cunningham et al. (2003) take this further, and give an elaborate presentation of the interactions, not only the social, cultural and linguistic factors that influence change in the society, but also the physical and external factors, such as worldview and influence of change agents, that play extremely significant roles to effect the change.

Communication is also influenced by the degree of 'homophily', or similarity, between the source of the message and the receiver. Homophily is defined by a sharing of important attributes that enhance the effectiveness of communication. Such attributes may be expressed broadly as lifestyles, beliefs, philosophies or idioms, all of which affect the encoding and decoding of messages. Coding refers to the manner in which a sender chooses to word and express a message. Decoding denotes the manner in which the receiver interprets the message (Buchanan and Huczynski 1991). The products of these twin processes are defined by 'perceptual filters', coloured by the context and setting in which the communication takes place, and refined by the relationship and mode of transmission between the sender and the receiver. In other words, these factors affect what people express outwardly and hold back, what they hear and do not hear, and the meaning they assign to what they hear.

Education for Sustainability

So far, we have gathered a substantial amount of information on the subject of environmental education. One of the conclusions from the knowledge gained is that the importance of this subject is increasingly enhanced by its role, both a prerequisite and a tool for sustainable development (Schneider and Weekes-Vagliani 1993). Its broad scope includes but is not limited to ecological and economic considerations. It is well known that as society strives to achieve economic development, through the use of environmental and other resources, ecological processes and life-support systems are often negatively affected. Negative effects on ecological processes and life-support systems are known to have serious consequences for the economic system and climate. Based on this realisation, there is now more and more emphasis on striking a balance between economic development and environmental protection; hence the concept of sustainable development.

The link between environmental education and sustainable development has been demonstrated by authors such as Huckle. In his 'Environmental education and sustainability: a view from critical theory', which appears as a chapter in Fien's book *Environmental Education: A Pathway to Sustainability*, published in 1993, he elaborates, on the discourses of 'education for environmental management and control' and 'education for sustainability'. The concept of education for sustainability has continued to grow in prominence. This is due, particularly, to its strong appeal emanating from its obvious focus on sustainable development. This can be discerned from its name, which soon made it gain wider acceptability as an alternative form of environmental education. It should, however, be noted that education for sustainability is a re-orientation of environmental education towards sustainable development considerations.

Therefore, while environmental education takes a holistic approach and maintains a balance between all the considerations, education for sustainability tends to lay more emphasis on sustainable development.

Conclusion

This chapter has established the fact that although environmental education is a relatively new subject in Africa, the concepts and efforts that eventually brought about its development into a full academic discipline have come a long way. It grew from warnings by concerned and foresighted individuals about the possible consequences of environmentally unsustainable activities perpetrated by humans, and developed into conservation and educational movements in the US in the 1930s.

The conservation movements focused on legislative and practical measures to safeguard nature and its resources; while the educational movements concentrated on the development of human knowledge about nature, through well designed studies. The latter movements have similarities with initiatives in the UK in the 1940s. However, the need for environmental education became more urgent in the 1960s due to the increasing evidence of the rapid degradation of the environment. Two international conferences were held, in 1972 and 1975 respectively. These helped formulate the definition and internationalisation of environmental education.

The efficacy of an environmental education programme lies in its inclusion of both short-term and long-term dimensions, meaning implementing both non-formal and formal aspects of the subject. We are advised of the necessity to ensure that the programme should, as much as possible, reflect local realities, colour and flavour.

It should be noted that implementing an environmental education programme requires, from the outset, an assessment of the environmental situation, an identification of the target audience and formulation of the working strategy. This is followed by a consideration of a range of factors that inform the formulation and communication of environmental messages. This chapter has also defined the meaning, scope and forms of environmental education.

Revision Questions

1. What important movements have given birth to what has come to be known as environmental education?
2. Discuss the role played by Aldo Leopold in the development of environmental education?
3. Discuss the factors that brought about the internationalisation of environmental education.
4. Define environmental education and state its objectives.
5. With illustrative examples, describe the three forms of environmental education.
6. Distinguish between awareness raising and sensitisation.
7. Discuss the four key prerequisites of environmental education.

Critical Thinking Questions

1. The scope of environmental education is more or less defined by the site and the available resources. Discuss.

2. In order to formulate appropriate strategies for changing or modifying the general attitude of a particular group of people towards the environment, a historical journey in time and space may be required. Discuss.
3. The more complex, less conspicuous or less compatible with existing practices an innovation is, the slower its diffusion rate. Elaborate on this within the context of a named environmental education activity.
4. The efficacy of an environmental education programme lies in its inclusion of both short-term and long-term dimensions. Discuss.
5. Discuss the extent to which environmental education merits the description of education for sustainability.

References

Adzeyuf, E. [no date], *Micro-media Communication Techniques. Training Module in Communication Skills for Human Resource Management and Development*, unpublished paper.

Boulton, M.N., and Knight, D., 1996, 'Conservation Education', in Spellberg, I.F., ed., *Conservation Biology*, London: Longman, pp. 69–79.

Buchanan, D. and Huczunski, A., 1991, *Organisational Behaviour: An Introductory Text*, Essex: Prentice Hall.

Chaffee, S.H., 1972, 'The Interpersonal Context of Mass Communication', in F.G., Klime, and P.J., Tichenor, eds., *Current Perspectives in Mass Communication Research*, Vol 1, Beverly Hills: Sage.

Cunningham, W.P., Saigo, B.W., and Cunningham, M.A., 2003, *Environmental Science: A Global Concern*, New York: McGraw-Hill.

Dunlop, J., 1993, 'Lessons from Environmental Education in Industrialised Countries', in H. Schneider, H., Vinke, J., and Weekes-Vagliani, W., eds, *Environmental Education: An Approach to Sustainable Development*, Paris: Organisation for Economic Cooperation and Development, pp. 79–101.

Horn, B.R. and Rogers, E.M., 1975, 'Interpersonal Communication of Environmental Education, in McInnis, N., and Albrecht, D., eds., *What Makes Education Environmental?* Washington DC: International Educators Inc.

Inyang, E., 2007, 'Baseline Study for the Development of an Environmental Education Programme in Korup National Park and Mount Cameroon Areas in Southwest Cameroon. A Consultancy Report Submitted to the WWF-Coastal Forests Programme', IUCN, 1993, *Education for Sustainability*, Gland: IUCN.

Martin, P., 1990, *First Steps to Sustainability: The School Curriculum and the Environment*, United Kingdom: WWF.

Ngome, M., 1992, *Environmental Education in Cameroon: The Problems and Prospects*, Cameroon: WWF.

Noibi, Y., 1990, *Pre-service Teacher Environmental Knowledge and Attitude: The Lagos Experience*, prepared for the 5th International Environmental Education Conference, Churchill School, Churchill, UK, 25 – 29 June 1990.

Patterson, T.E., 1993, *The American Democracy*, New York: McGraw-Hill.

Ritson, C., 1977, *Agricultural Economics: Principles and Policy*, London: Granada Publishing.

Rogers, E.M. and Shoemaker, F.F., 1971, 2nd edition, *Communication of Innovations: A Cross-cultural Approach*, New York: Free Press.

Rogers, E.M., 1973, 'Mass Media and Interpersonal Communication', in Pool, I.S. and Ashramm, W., eds., *Handbook of Communication*, Chicago: Rand.

Schneider, H., 1993, 'Conclusion', in H., Schneider, J., Vinke, and W., Weekes-Vagliani, eds., *Environmental Education: An Approach to Sustainable Development*, Paris: Organisation for Economic Co-operation and Development, pp147–7.

Schneider, H.J., Vinke and W. Weekes-Vagliani, eds., 1993, *Environmental Education: An Approach to Sustainable Development*, Paris: Organisation for Economic Co-operation and Development, pp. 79–101.

Seiler, W. J. and Beall, M.I., 2002, 5th edition, *Communication: Making Connections*, Boston: Allyn and Bacon.

Swan, M., 1975, 'Forerunners of Environmental Education', in N. McInnis, and D. Albrecht, eds., *What Makes Education Environmental?* Washington DC: InternationalEducators Inc.

The World Bank, 1992, *The World Development Report 1992*, Washington DC: Oxford University Press.

Touré, F. D., 1993, 'Environmental Awareness Raising and Education: Experience of the "Integrated Agro-Sylvo-Pastoral Development Project in Four Pilot Villages" in Senegal', in H., Schneider, J., Vinke and W., Weekes-Vagliani, eds., *Environmental Education: An Approach to Sustainable Development*, Paris: Organisation for Economic Co-operation and Development, pp. 147–7.

Vinke, J., 1993, 'Actors and Approaches in Environmental Education in Developing Countries', in H. Schneider, J. Vinke and W. Weekes-Vagliani, eds., *Environmental Education: An Approach to Sustainable Development*, Paris: Organisation for Economic Co-operation and Development, pp. 39–77.

Wheeler, K.S., 1981, 'A Brief History of Environmental Education in the UK', in *Department of Education and Science: A Review*, London: HMSO, pp. 22–3.

Wood, C., 1999, 'Environmental planning', in B. Cullingworth, ed., *British Planning: 50 Years of Urban and Regional Policy*, London: The Athlone Press.

Wood, S. and W. Wood, 1985, *Conservation Education: A Planning Guide. Peace Corps Information and Collection Exchange Manual M-23*, Washington DC: Training and Program Support.

WWF, 1988, *A Common Purpose: Environmental Education in the School Curriculum*, Surrey: WWF.

Chapter 10

Sustainability Reporting

Introduction

Sustainability reporting can be viewed at global, continental, sub-continental, national, sub-national and company level. The notions of sustainability and sustainable development are generally taken to reflect environmental, economic, social and technological spheres; whereas at the corporate level, these spheres are enveloped within the principles of good corporate governance. These include the elements of equity, accountability and transparency.

As an emerging phenomenon, sustainability reporting guidelines and initiatives are evolving at all levels. At a corporate level, sustainability reporting encompasses the now widely accepted concept of triple bottom line reporting; focusing on environmental, social, and economic sustainability, as well as good corporate governance. Since reporting at this level traditionally covered economic and/or financial matters adequately, this chapter seeks to highlight and elevate the other, previously neglected elements of reporting including environmental and social dimensions.

This chapter presents the fundamentals and applications of these two previously neglected dimensions in reporting and analyses how they have been addressed, particularly at national and company specific levels to sustain Africa's natural resource heritage. Therefore, the following five sections are critically examined in the chapter: global and national sustainability reporting initiatives; a historical perspective of sustainability in business; company level sustainability reporting guidelines; a global overview of sustainability reporting; and sustainability banking in Africa.

Global and National Sustainability Reporting Initiatives

This section considers global and national sustainability reporting guidelines and initiatives. Focus is on the following initiatives: the United Nations Indicators of Sustainable Development Framework of 2001; and Yale university's Environmental Sustainability Index and Environmental Performance indices where work started in 2000. Each of these initiatives will be considered briefly in turn in the next sections.

United Nations Indicator Framework

Work on the UN programme to set indicators for sustainable development was commissioned by the United Nations Commission of Sustainable Development (UNCSD) in 1995 (United Nations Division for Sustainable Development 2001). This resulted in the publication of a book, *Indicators of Sustainable Development: Guidelines and Methodologies*, in 2001. Its purpose was to stimulate and support further work, testing, and the developing of indicators at national government levels.

Indicators can provide important guidance for decision making in a number of ways (United Nations Division for Sustainable Development 2004). They can transform physical and social science knowledge into manageable units of information that permit informed decision making in environmental governance. Indicators also assist in measuring progress towards achieving stipulated sustainable development goals. They can provide early warnings where environmental disasters might occur, thereby preventing social and economic losses. No set of indicators can be final and definitive, and are adjusted over time.

As early as 1992, the UN's Agenda 21 called for the need for national governments to draw up indicators to measure their progress towards achieving sustainable development (UN 1992). The 2001 UN indicator framework for measuring sustainable development draws on four pillars of sustainability: social, environmental, economic and institutional (United Nations Division for Sustainable Development 2001), identified in the earlier work on indicators by the Commission on Sustainable Development (United Nations Division for Sustainable Development 1999). Such indicators were grouped as the driving force, state and response characteristics. Driving force denotes human activities, processes and patterns that impact on, either positively or negatively, and shape, sustainable futures. The state indicators give a measure on the condition of sustainable development. Response indicators represent societal actions targeted at moving towards achieving sustainability in various sectors (United Nations Division for Sustainable Development 2001). Although the themes and sub-themes from the indicators were designed to guide national governments, the first three pillars have been widely used to develop themes and sub-themes for company level sustainability indicators and reporting initiatives.

One hundred and thirty-four indicators were developed between 1996 and 1999, and administered on a voluntary basis to twenty-two countries (United Nations Division for Sustainable Development 1999). The pilot-testing country feedbacks and subsequent work on the indicator framework suggested forty-six key thematic indicators across the four pillars mentioned earlier (Table 10.1).

Table 10.1: Key Thematic Indicator Areas

Pillar	Key thematic indicators
Social	• Education • Employment • Health/water supply/sanitation • Hounsing • Welfare and quality of life • Cultural heritage • Poverty/income distribution • Crime • Population • Social and ethical values • Role of women • Access to land resources • Community structure • Equity/social exclusion
Environmental	• Freshwater/groundwater • Agriculture/secure food supply • Urban • Coastal zone • Marine environmental/coral reef protection • Fisheries • Biodiversity/biotechnology • Sustainable forest management • Air pollution and ozone depletion • Global climate change/sea level rise • Sustainable use of natural resources • Sustainable tourism • Resctricted carrying capacity • Land use change
Economic	• Economic dependency/indebtedness/ODA • Energy • Consumption and production patterns • Transportation • Mining • Economic structure and development • Trade • Productivity
Institutional	• Integrated decision-making • Capacity building • Science and technology

- Public awareness and information
- International conventions and cooperation
- Governance/role of civil society
- Institutional and legislative frameworks
- Disaster preparedness
- Public participation

Source: UN Department of Economic and Social Affairs (2001: 22).

From the key indicator themes, sub-themes and measurable indicators are drawn. For example, taking land as an environmental theme, the sub-themes would include agriculture, forest, desertification and urbanisation. Under agriculture, the sub-themes measures would be arable and permanent crop land area, use of fertilisers and use of agricultural pesticides. From the forest sub-theme, we would measure forest area as percentage of the total land area, as well as wood harvesting intensity. This is the kind of framework that the UN sustainable development indicators seeks to achieve. The framework has also been adopted and applied in specific sectors to measure sustainability. This has been the case with its new application within the energy sector by the International Atomic Energy Agency (IAEA), as briefly outlined in the following section.

In 2005 the IAEA, in collaboration with the UN Department of Economic and Social Affairs, the International Energy Agency, Eurosat, and the European Environmental Agency, published work on energy indicators (International Atomic Energy Agency 2005). The work provides guidelines and methodologies along the lines of the UN framework of indicators for sustainable development, but focuses on three pillars: the social, the environmental and the economic. The core set of Energy Indicators for Sustainable Development (EISD) provides information on current energy-related trends in a form that assists decision making at a national level. Thus nations are aided in assessing the effectiveness of policies for action towards achieving sustainability in the energy sector. These energy indicators provide benchmarks for the WSSD Implementation Plan energy targets, which include the need to:

- integrate energy into socio-economic programmes
- combine more renewable energy, energy efficiency and advanced energy technologies to meet the growing need for energy services
- increase the share of renewable energy options
- reduce the flaring and venting of gas
- establish domestic programmes on energy efficiency
- improve the functioning and transparency of information in energy markets
- reduce market distortions and assist developing countries in their domestic efforts to provide energy services to all sectors of their populations (International Atomic Energy Agency 2005: 6; UN, 2002).

The energy indicators for sustainability incorporate social sub-themes such as accessibility, affordability, disparities and safety. Some of the indicators include the share of households (or population) without electricity and the share of household income spent on fuel and electricity. The economic sub-themes include overall use, overall productivity, supply efficiency, production, end use, diversification (fuel mix), prices, imports and strategic fuel stocks. Some of the indicators include per capita energy-use, and per unit of gross domestic product (GDP), and the ratio of resources to production. The environment sub-themes cover climate change, air quality, water quality, soil quality, forest as well as solid waste generation and management. Some of the indicators include greenhouse gas emissions from energy production and use per capita and per unit of GDP, and the ratio of solid waste generation to units of energy produced. The energy indicators continue to be modified as per the dictates of new technology and information regarding energy. Readers are encouraged to trace new work and references.

Yale University Environmental Sustainability and Performance Indices

Work on Environmental Sustainability Index (ESI) and Environmental Performance Index (EPI) by Yale University and its collaborating partners started in 2000. It resulted in the publication of ESIs and EPIs in 2000, 2001, 2002, 2005 and 2006. For our purposes, the indices for the three most recent years are discussed.

In 2002 a pilot EPI was designed to measure environmental stewardship at a national level. Four core indicators were used and measured: air quality, water quality, greenhouse gas emissions and land protection/degradation. Only seven African countries appeared in the top fifty countries, out of the 142 countries assessed. Botswana was ranked thirteenth, Namibia, twenty-sixth, Zimbabwe, forty-sixth, and South Africa, seventy-seventh (Esty, Levy et al. 2005).

The 2005 ESI improved on the previous versions. This work was still directed by the Yale University's Centre for Environmental Law and Policy and the Centre for International Earth Science Information Network of Columbia University in collaboration with the World Economic Forum and the Joint Research Centre of the European Commission (Esty, Levy et al. 2005). The 2005 ESI aimed at benchmarking national environmental stewardship for the next decades. The 2005 ESI benchmarks countries by integrating seventy-six data sets that trace natural resource endowments, past and current pollution levels, environmental management efforts and the capacity of nations to improve their environmental performance. The seventy-six data sets were aggregated into twenty-one indicators of environmental sustainability. From the authors, the twenty-one indicators allow direct comparison across five broad themes: environmental systems; reducing environmental stress; reducing human vulnerability to environmental stresses; societal and institutional capacity to respond to environmental challenges; and global stewardship.

The higher a nation's ESI score, the better its chances of maintaining favourable environmental conditions in the future. Interestingly, out of the 146 countries included in the 2005 ESI, none of the forty African countries came in the top ten. Africa's top country, Gabon, was ranked twelfth, and was the only one in the top twenty. In terms of African countries ranked, the top ten were: Gabon, followed by Central African Republic, Namibia, Botswana, Mali, Ghana, Cameroon, Tunisia, Uganda and Senegal. The bottom ten were: Egypt, followed by Sierra Leone, Liberia, Angola, Mauritania, Libya, Zimbabwe, Burundi, Ethiopia and Sudan.

Work on the ESI continued in 2006, resulting in another publication on Environmental Performance Index (EPI). EPI is pivoted on twin broad environmental protection objectives: (1) to reduce environmental stresses on human health, and (2) to promote the vitality and sound natural resource management of ecosystems (Esty et al. 2006: 1). These objectives were formulated to address, in particular, some of the concerns that were raised against the lack of measurability of the environmental objectives of the Millennium Development Goals. The 2006 Pilot EPI, as it is commonly known, measured environmental health and ecosystems viability using sixteen indicators grouped into six well-established policy categories recorded by the authors as: environmental health, air quality, water resources, productive natural resources, biodiversity and habitat, and sustainable energy.

The top five nations from the EPI, in a survey of 133 countries, were New Zealand, Sweden, Finland, the Czech Republic and the UK. The lowest ranked countries were Ethiopia, Mali, Mauritania, Chad and Niger (Esty et al. 2006), all African countries. Indicators across the six major categories, cited above, included child mortality, indoor air pollution, drinking water, adequate sanitation, urban particulate matter, regional ozone, nitrogen loading, water consumption, wilderness protection, timber harvest rates, agricultural subsidies, over-fishing, energy efficiency, renewable energy and carbon dioxide per GDP. Given the focus of this book on environmental management in Africa, the entire rankings and scores for thirty-eight AU member countries, involved in the survey, are presented in Table 10.2.

Table 10.2: EPI for AU Countries, 2006

Rank	Country	Score	Rank	Country	Score	Rank	Country	Score
1	Gabon	73.2	14	Malawi	56.5	27	Guinea	49.2
2	Algeria	66.2	15	Namibia	56.5	28	Madagascar	48.5
3	Ghana	63.1	16	Kenya	56.4	29	Guinea-Bissau	46.1
4	Zimbabwe	63.0	17	Zambia	54.4	30	Mozambique	45.7
5	South Africa	62.0	18	Cameroon	54.1	31	Nigeria	44.5
6	Uganda	60.8	19	Swaziland	53.9	32	Sudan	44.0
7	Tunisia	60.0	20	Togo	52.8	33	Burkina Faso	43.2
8	Tanzania	59.0	21	Gambia	52.3	34	Angola	39.3
9	Benin	58.4	22	Senegal	52.1	35	Mali	36.7
10	Egypt	57.9	23	Burundi	51.6	36	Mauritania	33.9
11	Ivory Coast	57.5	24	Liberia	51.0	37	Chad	32.0
12	Central Afr.	57.3	25	Sierra Leone	49.5	38	Niger	30.5
13	Rwanda	57.0	26	Congo	49.4			

Source: (Esty et al. 2006:19).

The authors note that all the top global performing nations invested heavily in protecting the environmental health of their citizens. However, of concern is that Africa's highest ranked, Gabon is ranked forty-sixth globally. The other top five are ranked sixty-third (Algeria), seventy-second (Ghana), seventy-fourth (Zimbabwe) and seventy-sixth (South Africa). Such statistics show that we, as a continent, still have much to do in terms of the EPI and sustainability reporting.

Sustainability in Business: A Historical Perspective

During the 1960s and 1970s the world, including corporations, continually denied their negative impacts on the environment. However, a series of severe and visible environmental disasters, such as the death of Lake Erie in the US, the Rhine river in Europe, and people dying of mercury poisoning in Japan forced a change of mindset (Hart 2004). Sustainability issues in business and industry were placed on the global agenda in the mid-1990s (Timberlake 1992). This emerged from an initiative by the UN Conference on Environment and Development Secretary-General aimed at raising environmental awareness to businesses during the Earth Summit of 1992. The result was the establishment of the Business Council for Sustainable Development (BCSD), and the subsequent publication of a book entitled *Changing Course: A Global Business Perspective on Development and the Environment*. The book became part of the Rio summit proceedings. It was published in seven languages prior to the Rio summit. The BCSD was made up of representatives from chief executives from Europe, North and South America, Asia, Africa and Australia.

According to Timberlake (1992:29), eco-efficiency was agreed on as the key feature of future sustainable businesses. Eco-efficiency was described as the 'production of goods and services whilst reducing resource consumption and pollution'. The eco-efficiency principle is drawn from of Principle 8 of the Rio Declaration, stipulating that to achieve sustainable development and a higher quality of life for all people, states should reduce and eliminate unsustainable patterns of production and consumption, and promote appropriate demographic policies (UN 1992).

To raise awareness of the need to integrate developmental issues and the environment, the principles of eco-efficiency were promoted through the BCSD book in twenty countries, particularly to those in the developed world. The BCSD unanimously agreed that the future winners in business will embark on improving their eco-efficiency because:

- customers were now demanding cleaner products and services;
- insurance companies were becoming more amenable to covering clean companies;
- employees, especially the best and the brightest were preferring to work for environmentally responsible entities;
- environmental regulations were getting tougher and would continue doing so in the future;

- new economic instruments such as taxes, charges and trade permits were rewarding clean companies; banks were more willing to finance companies that conserve the environment and prevent pollution rather than having to pay for clean-ups (Timberlake 1992).

The last point was basically a wake up call to businesses, geared towards end-of-pipe measures in managing the environment, to move over to pro-active, anticipatory entities when dealing with environmental concerns. Overall, eco-efficiency was deemed to help, rather than hurt, profitability.

Company Level Sustainability Reporting Guidelines

Many companies now use the Global Reporting Initiative's (GRI) Sustainability Reporting Guidelines (SRG), when reporting on sustainability issues. The GRI was launched in 1997 as a joint initiative of the US NGO Coalition for Environmentally Responsible Economies and the United Nations Environment Programme (UNEP). Its major goal was to enhance the equality, rigour and utility of sustainability reporting globally (Global Reporting Initiative 2002). By 1999, the first GRI SRG document was ready. The second was published in 2000. Currently, the 2002 edition is being used, and over 500 organisations, world-wide, were reported to have been using the guidelines by September 2004 (ACCA 2004).

The GRI initiative of 2002 provides twin sets of indicators: core and additional indicators (Global Reporting Initiative 2002). The core indictors are relevant to most reporting organisations and their key stakeholders; whilst the later may be concerned with issues such as leading practices in environmental, social and economic measurement.

The sustainability reporting guidelines are divided into four major parts. Part A covers aspects pertaining to using the GRI guidelines, Part B focuses on the reporting principles, Part C, looks at the report content, and Part D, comprises the glossary and annexes. The contents of the parts under review will each be considered briefly in turn.

The reporting principles focus on transparency, inclusiveness, auditability, completeness, relevance, sustainability context, accuracy, neutrality, comparability, clarity and timeliness (Box 10.1).

Box 10.1: GRI Sustainability Reporting Principles

P1-Transparency: Is the report providing full disclosure of the processes, procedures, and assumptions in report preparation?

P2-Inclusiveness: Is the reporting organisation systematically engaging its stakeholders to help focus and continually enhance the quality of its reports?

P3-Auditability: Is the reported data and information recorded, compiled, analysed, and disclosed in a way that would enable internal auditors or external assurance providers to attest to its reliability?

P4-Completeness: Is the report including all information that is material to users for assessing the organisation's economic, environmental, and social performance in a manner consistent with the declared boundaries, scope, and time period?

P5-Relevance: Is the report clearly defining the degree of importance assigned to particular indicators, including the threshold at which the information becomes significant enough to be reported?

P6-Sustainability Context: Is the report providing an overview of the context in which the data is reported relative to the larger ecological, social or economic constraints?

P7-Accuracy: Is the report achieving a high degree of exactness, or a low margin of error, such that users can make decisions with a high degree of confidence?

P8-Neutrality: Is the report avoiding bias in selection and presentation of information, and provide a balanced account of the organisation's performance?

P9-Comparability: Is the report maintaining consistency with previous reports in the boundary and scope of indicators? Alternatively, are any changes of boundary or scope, or re-statements of previously disclosed information, adequately disclosed?

P10-Clarity: Is the report making the reported information available in a manner that is responsive to the maximum number of users while still maintaining a suitable level of detail?

P11-Timeliness: Is the report being released in a manner that is consistent with a regular schedule that meets user needs?

Source: Modified after Global Reporting Initiative (2002:23–31).

Table 10.3: Suggested Categories of Reporting Indicators

	Category	Aspect
Economic	Direct economic impacts	• Customers • Suppliers • Employers • Providers of capital • Public sector • Materials • Energy • Water • Biodiversity
Environmental	Envionmental	• Emissions, effluent and waste • Suppliers • Products and services • Compliance • Transport • Overall • Employment
	Labour practice and decent work	• Labour/management relations • Health and safety • Training and education
Social	Human rights	• Diversity and opportunity • Strategy and management • Non-discrimination • Freedom of association and collective bargaining
	Society	• Child labour • Forced and compulsory labour • Disciplinary practices • Security practices • Indigenous rights • Community • Bribery and corrutpion • Political conditions • Competition and pricing
	Product responsibility	• Custom health and safety • Products and services • Advertising • Respect for privacy

Source: Global Reporting Initiative (2002:36).

The report content captures variables such as vision and strategy, profile, governance structure and management systems, the GRI content index and performance indicators. The performance indicators include the pillars of sustainability reporting that embrace economic, environmental and social sustainability. The annexes included are: an overview of the GRI, linkages between sustainability and financial reporting, guidance on incremental application of the GRI SRG, credibility and assurance, GRI indicators, and the GRI content index. Further details regarding elements discussed above and more on sustainability reporting can be obtained from the main document and Table 10.3 gives a summary of indicators recommended by the GRI.

Corporate responsibility has become a major issue in Europe. Some countries have passed legislation to this effect, or are being pressed to do so by lobby groups. Some of the recent bills on corporate responsibility from European countries include on social labelling (Belgium), sustainability reporting (Netherlands and France) and pension fund disclosure (the UK and Germany).

Sustainability Reporting: A Global Overview

Since publication of *Our Common Future*, the trend for better corporate governance and accountability has placed emphasis on the responsibilities of organisations towards all stakeholders, the environment and societies in which business is conducted (ACCA 2004). The practice of sustainability reporting has emerged from this doubled quest for greater organisational transparency. Since the concept of sustainability reporting is relatively new, selected definitions for some of the fundamental terms are provided here. An 'indicator' refers to a measure that can either be qualitative or quantitative in nature. Such measures are used to compute an index or indices. An 'index' is usually denoted as a single value that takes into consideration different weightings and aggregates from the indicators. 'Baselines' and 'benchmarks' are other terms. A baseline should be taken as the starting point. For example, since many companies were not reporting on sustainability issues earlier on, their very first reports are baseline reports. A benchmark is a standard against which organisations can assess compliance. For example, set national effluent standards for a pulp and paper mill are used as benchmarks before effluence is disposed of in a public water system or municipal sewer network. Jonah and Pienaar (2004) have developed additional concepts, adopted here for clarity. Corporate social responsibility (CSR) focuses on the socio-economic, ethical and moral responsibilities set in response to the changes and demands of society at large. Corporate social investment (CSI) includes the funding of and involvement in socio-economic upliftment. But it excludes employee benefits. Examples include education, housing, health, welfare, job creation, community development or empowerment. Corporate citizenship includes accountability for social, environmental and economic impact; engagement with stakeholders; and integration into mainstream business. Although the concept of sustainability reporting is still in infancy, in many African countries, the practice was first recorded in the early 1990s. In 1993, fewer than 100 sustainability reports were produced. By 1999, the figure increased more than five-fold. In 2003, over 1,500 reports were recorded globally (Jonah and Pienaar 2004). A comparison of sustainability reports produced on the global scale is presented in Graph 10.1.

Graph 10.1: Reports Produced, 1990–2003 (n=6,619)

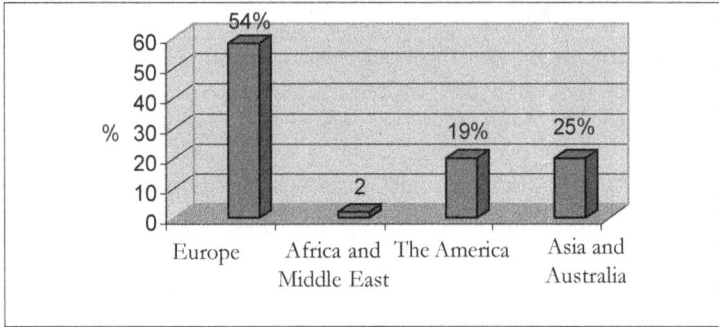

Source: Compiled from ACCA (2004:9).

The types and formats of the sampled reports produced between 2001 and 2003 (n=3,637) varied (ACCA 2004). The list and proportions were distributed as: sustainability reports (14 per cent), corporate responsibility (8%), annual with substantive non-financial sections (6%), community (3%), social (5%), environment and social (8%), environment, health and safety (14%) and environment (42%). From 3,637 reports produced between 2001 and 2003, Europe had the largest share, with 54 per cent, followed by Asia and Australasia with 25 per cent. The Americas had 19 per cent, and lastly Africa and the Middle East had a mere 2 per cent.

Given the growing concern and need for companies to report on non-financial matters, ACCA launched the world's first environmental reporting awards in 1990. The awards opened up spaces for engagement within corporate accountability. The awards achieved a number of things. They:

• highlighted the business community's role in sustainable development

• raised awareness and understanding of environmental reporting issues, and promoted the need for this type of discourse

• demonstrated the need for all business to be accountable for all their impacts on society

• showed that it was not just the shareholders who were interested in corporate activity, but that other stakeholders, too, had information needs (which were not being met)

• encouraged a number of organizations to prepare environmental reporting guidance material

• ultimately, by rewarding best practice and providing feedback via judges' reports, they helped to improve both the quality and the quantity of reporting we see in the world today (ACCA 2004:11).

Since 1990, mandatory environmental reporting has been introduced in some countries: Denmark, France, the Netherlands, Norway and Hong Kong. In Africa, South Africa is the sole country to have worked out a code of corporate governance ethics, and the only country in the world to have mandatory sustainability reporting for listed companies (African Institute of Corporate Citizenship 2004; UNEP FI 2005).

Sustainability Reporting in Africa

The first sustainability reports appeared in 1993, with external assurances first appearing in 1998 (ACCA 2004). Since 2002, there has been an exponential growth in reporting. This is due to many reports emerging from South Africa, according to the King II Report (Wixley and Everingham 2002; IDSA 2002) (for further details, see below), and the Johannesburg Stock Exchange requirements (Johannesburg Stock Exchange 2003). Out of the ninety-seven sustainability reports sampled for Africa, South Africa accounted for 75 per cent (ACCA 2004). The top five typology of the sustainability reports were ranked as: sustainability; environment and social; environment, health and safety; corporate responsibility and social. The remaining percentage was shared between Algeria, Gabon, Mauritius, Mozambique, Namibia, Nigeria, Uganda and Zimbabwe.

The Sustainability Reporting Nexus

The nexus between corporate governance, the pillars of sustainable development and triple bottom line (or sustainability) reporting, technological advancement and a sustainable organisation should already have been grasped. Various models can be used to depict these linkages. Figure 10.2 is an example.

Figure 10.2: Sustainability Reporting Nexus

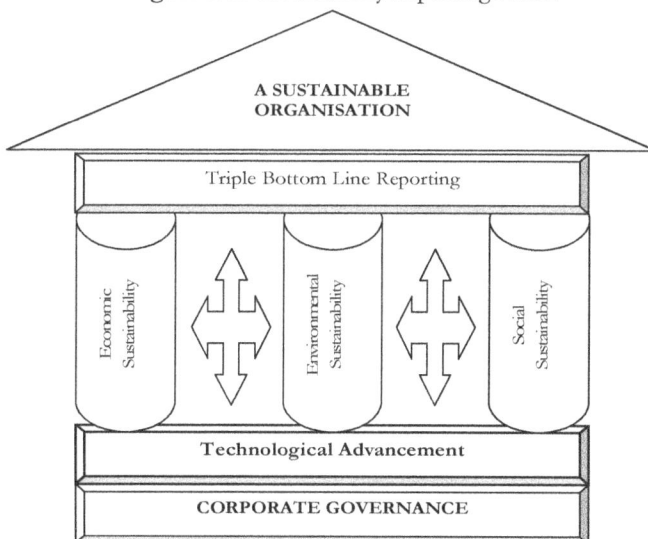

Figure 10.2 reveals that, like a real house roof, a sustainable organisation rests upon the strengths of the three pillars of sustainability: the economic, the environmental and the social. In addition to the pillars, African organisations, or businesses, that will survive into the future must have their roots founded on firm ground, built on technology, particularly research and development, and the principles of corporate governance. The four directional arrows within the figure indicate the many multi-dimensional relationships and the nexus that emerges in the realm of corporate governance and sustainability reporting. Further deliberations concerning the concepts introduced above are made within this chapter.

Understanding the 'Triple Bottom Line' Metaphor

The triple bottom line (TBL) of sustainable development seeks to account for economic prosperity, environmental (ecological and/or ecosystems) quality, and social justice (Elkington 2004). Environmental quality and social justice have been neglected by business and industry for a long time. Each of the three pillars can be discussed in their relationship to accountability, accounting, auditing and reporting. Traditionally, a company's bottom-line is associated with the profit figure, resulting from the deductions of cost and depreciation of capital. This is part of standard accounting practice. Hence in line with the TBL concept, there should be an equal accounting (the pulling together, recording and analysis) of a wide range of environmental and social data, including figures.

Hart (2004) maintains that pollution, depletion of natural resources and poverty are the key challenges to attaining sustainability. These key challenges are manifest in varying degrees in countries from the developed economies, emerging economies and what he refers to as the survival economies. Many African countries still fall within the survival category. Many others, that had shown signs of becoming emerging economies, are, for various reasons, degenerating back into survival economies. Table 10.4 summaries some of the challenges to sustainability.

Table 10.4: Sustainability Challenges

Taxonomy	Pollution	Depletion of natural resources	Poverty
Developed economies	• Greenhouse gases • Use of toxic materials • Contaminated sites	• Scarcity of materials • Insufficient reuse and recycling	• Urban and minority unemployment
Emerging economies	• Industrial emissions • Contaminated water • Lack of sewage treatment	• Overexploitation of renewable resources • Overuse of water for irrigation	• Migration to cities • Lack of skilled workers • Income inequality
Survival economies	• Dung and wood burning • Lack of sanitation • Ecosystems destruction due to development	• Deforestation • Overgrazing • Soil loss	• Population growth • Low status of women • dislocation

Source: Hart (2004:11).

Hart (2004) also identifies three key strategies to overcome the problems associated with unsustainable behaviour, globally. He outlines sequential stages that start with the prevention of pollution (RSA 1998), the institution of product stewardship (Fishbein 1994) and clean technology (Pauli 1997). These three strategies can be integrated into what he terms as a 'sustainability portfolio', which examines company issues of the day and the future, from the internal and external perspectives. The green, or environmental, bottom-line becomes paramount. The green bottom-line requires that environment-related management accounting be undertaken (Bennett and James 2004). The environmental bottom-line mainly applies at company level, starting with the organisation itself, and progresses to its supply chain and relationship with the community. Accounting in this respect should focus on both financial and non-financial issues, within this specific bottom-line. Six domains are visible when dealing with the green-bottom-line (Table 10.5).

Table 10.5: Company Level Environment-related Management Accounting

Scale/Focus	Organisation	Supply chain	Society
Financial Focus	Environment-related Financial Management	Life-cycle Cost Assessment	Environmental externalities Costing
Non-financial Focus	Energy and Materials Accounting	Life-cycle Assessment	Environmental Impact Assessment

Source: Bennet and James (2004:127).

Such environment-related management accounting is carried out to inform and support decision making processes, influenced by environmental factors. Some of the main objectives of this form of accounting include the need to: demonstrate the impact on the income statement of environment-related activities; prioritise environmental actions; enhance customer value; and support sustainable business (Bennett et al. 2004). Overall, the triple bottom line aims to achieve:

- transparency and effectiveness: allowing people to assess or ensure that organisations are doing the right thing in terms of their core business;

- accountability: allowing organizations to take responsibility for their actions and to report this honestly to their stakeholders;

- consultation and responsiveness: enabling organisations to ensure positive relationships both internally and externally and responding to the feedback from stakeholders through informed and appropriate decision making;

- impact assessment: allowing organisations to identify the nature and scope of impact of actions they take particularly across and between the three bottom lines;

- information and communication (including public relations): enabling organisations to use the results of their processes for future decision making and to convey, as and when appropriate, these results to the public (Mahoney and Potter 2004:154).

The Triple Bottom Line in South Africa

The triple bottom line reporting (TBLR) concept in South Africa calls for integrated sustainability reporting (ISR) in the corporate world. Sustainability from the King II Report (IDSA 2002) entails a focus on previously sidelined non-financial aspects of corporate practice, which have been found to influence the ability of an enterprise to survive and prosper in the communities in which it conducts its business. The King II Report noted that many companies in South Africa were not reporting comprehensively on the environment. The King II Report recommended that the environment be considered as a stakeholder in its own right. Where a company operates in a foreign land, where higher environmental standards apply, these standards should be implemented in South Africa. Meanwhile, the 'best practice environmental option', which has most benefits, or causes least damage, to the environment at a cost accepted by the society, should be applied to all decisions taken by a business entity.

For South African companies, TBL reporting became effective for the financial years beginning on or after 1 March 2002. It is applicable to listed companies, banks, financial and insurance entities and certain public sector entities (Wixley et al. 2002). Although it is a voluntary practice, the Johannesburg Stock Exchange (JSE) requires that all listed companies adhere to the TBL reporting system. In addition, the Code of Corporate Practice and Conduct, spelt out by the King II Report, contains a clause that places a duty and responsibility on all boards and individual directors to make sure that its provisions are adhered to.

The King II Report deals with Corporate Governance issues (IDSA 1994; IDSA 2002) and outlines a Code of Corporate Practice and Conduct (Wixley et al. 2002) for South African business and industry. The King II Report is made up of six major sections that deal with: boards and directors, risk management, internal audit, integrated sustainability reporting (key focus area), accounting and auditing, and compliance and enforcement.

In 2002, Ernest and Young started incorporating sustainability reporting aspects into the adjudication process, as part of its Excellence in Corporate Reporting (ECR) survey initiative. The ECR survey initiative, pioneered in 1997, originally focused on encouraging excellence in the quality of financial reporting by South Africa's top 100 companies to investors and other stakeholders (Ernest and Young 2004). The shift in emphasis from excellence in financial reporting to excellence in corporate reporting has been necessitated by the need to evaluate company annual reports from the perspective of the broad community of stakeholders. The 2003/4 adjudication reflected, for the first time, the response to the King II Report on corporate governance and sustainability reporting.

Potentially 400 points can be obtained by companies, the annual reports, of which, are assessed. These points are given the following weighting: performance review (approximate weighting, 35%), financial disclosure (35%), forward-looking information (20%), and presentation (10%). It is important to note that elements related to corporate governance and sustainability disclosures fall within the performance review category. Further details and free copies of the 2003/4 and past ECR survey initiatives can be obtained from www.ey.com/za.

There has been a growing commitment by South African companies to fulfil corporate social responsibility requirements. Companies cannot account for profitability alone, without taking cognisance of CSR. A corollary of the rising interest in CSR has been the growth of socially responsible investment (SRI). Investors are becoming attracted to CSR as a business approach, because it creates long-term shareholder value by embracing opportunities, maximising efficiencies and managing risks derived from economic, environmental and social developments that are not necessarily addressed by short-term financial analysis. This reflects the so-called triple bottom line approach.

To this effect, the FTSE/JSE have come up with a SRI index as a benchmark to facilitate investment in companies with good records of CSR. The SRI Index, drawing heavily on the GRI SRG, is to be constituted from companies that form part of the FTSE/JSE All Share Index, and which meet the selection criteria set out in the SRI Index Philosophy and Criteria (Johannesburg Stock Exchange 2003; Johannesburg Stock Exchange 2004). The selection criteria, meeting both local and international requirements, cover three areas of principle: environmental sustainability; positive relationships with stakeholders; and upholding and supporting universal human rights.

In order to assess listed companies against the JSE SRI Index, in October 2003, the JSE circulated a 56-page launch questionnaire investigating major SRI issues. In addition to filling in the questionnaire, companies were requested to provide additional information or documentation, such as annual reports, the text of relevant policies, brochures to customers, or any other communication to stakeholders or the general public (Johannesburg Stock Exchange 2004). From the questionnaire, the following major sub-themes of TBLR were identified.

Economic Sustainability

Covers the sub-themes of policies (generic for all the three pillars), governance and management (generic for all the three pillars), ownership of the company, salaries and remuneration, knowledge management, human resources (generic for all the three pillars), contractors (generic for all the three pillars), reporting, auditing and accounting (generic for all the three pillars), insurance and contingency plans, customers and products and compliance.

Environmental Sustainability

This includes sub-themes such as impact assessment, environmental management systems, biodiversity, natural resources and genetically modified organisms, emissions and discharges, energy, waste, water, accidents and incidents, auditing, accounting and reporting, compliance, standards and certification/de-certification and awards.

Social Sustainability

Addressing the following sub-themes: black economic empowerment (BEE), health and safety, HIV/Aids and other chronic occupational diseases, human rights, community relations, corporate social investment and awards. Social questions also covered, in much depth, parameters such as corporate governance, ethics, corruption, bribery and money laundering, stakeholder engagement, BEE, human resources (including skills development), health and safety, HIV/Aids, employment equity, diversification and transformation, human rights, community development, and consumer groups. In order to be included in the JSE SRI Index, companies had to achieve an overall score of at least seventy points, and surpass individual thresholds that varied from category to category, for high impact, medium impact or low impact entities. The scoring method is based on the extent to which a company adopts and/or implements the four sustainability pillars indicators of corporate governance, environmental sustainability, economic sustainability and social sustainability as follows:

- None: Nothing in place and only sporadic or ad hoc activity takes palace, if any (score of 0)

- Partial or efforts: Objectives/systems are in place, but do not meet the level set by the Criteria; or evidence exists that regular/systematic efforts are being made to set objectives/implement a system (score of 1)

- Full or complete: Objectives/systems are in place and are reported on, fully meeting the level set by the Criteria (score of 2)

- Exceeding: Objectives/systems are in place exceeding the level set by the Criteria (score of 3) (Johannesburg Stock Exchange 2005:5).

However, participants of the round-table discussions highlighted a number of concerns, including: a request for more definitional clarity on issues surrounding sustainability reporting, and the lack of expertise within organisations around non-financial reporting. The outcome of the 2003 SRI index is represented in Box 10.2.

Box 10.2: Press Release – Launch of JSE Socially Responsible Investment Index

---

19 May 2004

The JSE Securities Exchange South Africa today announced the names of the constituents of the JSE's first Socially Responsible Investment Index ('SRI Index'). Calculation of the Index will commence tomorrow.

JSE Deputy CEO, Nicky Newton-King said: 'The last few years have seen an increasing awareness of and need to measure sustainable business practices. In South Africa, in particular, the Second King Report on Corporate Governance urges companies to embrace the triple bottom line as a method of doing business. The JSE has been working with people across society's spectrum as well as the JSE SRI Advisory Committee to create the SRI Index as a means of helping to focus the debate on triple bottom line practices, in addition to recognising the tremendous efforts already made by South African companies in this area.'

The 'SRI Index measures companies' policies, performance and reporting in relation to the three pillars of the tripple bottom line (environmental, econmic and social sustainability), as well as corporate governance practice.

All the companies in the FTSE/JSE All Share Index were invited to participate in the process which lead to the JSE SRI Index on a voluntary basis. 74 listed companies participated and 51 companies met the criteria. Their names are attached to this release in alphabetical order. The SRI Index is the first index of this nature in an emerging market and the first in the world to be launched by an exchange. Newton-King said that 'the JSE is delighted that so many listed companies participated in the process and that 51 companies are now part of the first SRI Index. A notable aspect of the constituents of the SRI Index is that 17 of the constituents are part of our MidCap Index and 3 are part of the SmallCap Index which reflects the fact that sustainability is a business issue for companies of all sizes.'

Data collection and analysis was done by Sustainability Research & Intelligence, and KPMG performed independent assurance of the analysis process.

In launching the SRI Index, the JSE announced that Graca Machel and Reuel Khoza had agreed to be the Patrons of the SRI Index. Newton-King said the JSE was honoured by their involvement which was very important for the JSE.

Constituents of the first JSE SRI Index (in alphabetical order)

ABSA Group Ltd
African Bank Investments Ltd
African Oxygen Ltd
African Rainbow Minerals Ltd
Alexander Forbes Ltd
Allied Electronics Corporation Ltd
Allied Technologies Ltd
Amalgamated Beverage Industries Ltd
Anglo American Platinum Corporation Ltd
Anglo American plc
Anglogold Ashanti Ltd
Aveng Ltd
AVI Ltd
Barloworld Ltd
BHP Billiton plc
The Bidvest Group Ltd
City Lodge Hotels Ltd
Dimension Data Holdings plc
Edgars Consolidated Stores Ltd
FirstRand Ltd
Gold Fields Ltd

Iscor Ltd
Johnnic Communications Ltd
Kumba Resources Ltd
Liberty International Plc
Massmart Holdings Ltd
Medi-clinic Corporation Ltd
MTN Group Ltd
Murray and Roberts Holdings Ltd
Nampak Ltd
Nedcor Ltd
Network Healthcare Holdings Ltd
Northam Platinum Ltd
Old Mutual plc
Pick n Pay Holdings Ltd
Pretoria Portland Cement Company Ltd
Remgro Ltd
SABMiller plc
Sappi Ltd
Sasol Ltd
South African Chrome and Alloys Ltd
Standard Bank Group Ltd

Gold Reef Casino Resorts Ltd	Telkom SA Ltd
Harmony Gold Mining Company Ltd	The Tongaat-Hulett Group Ltd
Illovo Sugar Ltd	Venfin Ltd
Impala Platinum Holdings Ltd	Woolworths Holdings Ltd
Investec Ltd	

Source: http://www.jse.co.za/news/sri_launch.doc, accessed 1 April 2005.

The SRI index criteria have changed since its launch in 2003. The overall aggregate points have been increased from 70 to 78 for the 2004 and 2005 reporting periods respectively. Similarly, the other criteria have been adjusted upwards with corporate governance moving from 12 to 16 points; social sustainability from 22 to 25 points; and economic sustainability from 18 to 21 points (Graph 10.3).

Graph 10.3: Environmental Sustainability Criteria, 2003–2005

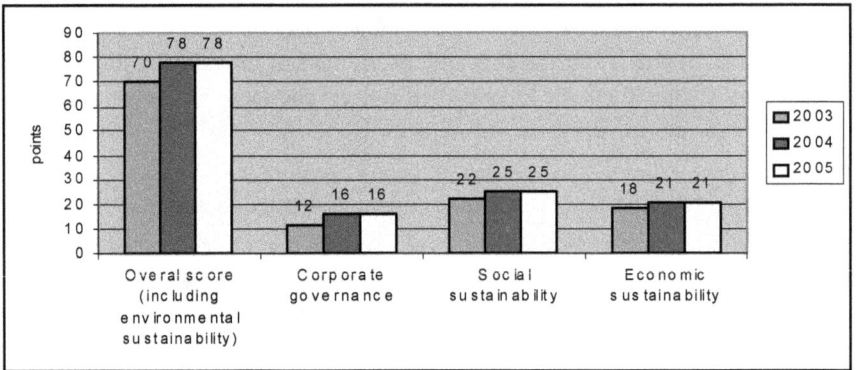

The criteria for environmental sustainability were also adjusted upwards. In 2004 and 2005, low impact companies were expected to score 9 points, instead of the original 8 points, established in 2003. Medium impact companies were expected to achieve a score of 16, up from 14, and high impact companies had to garner 22 points, instead of 20 (Graph 10.4).

Graph 10.4: Environmental Criteria Points

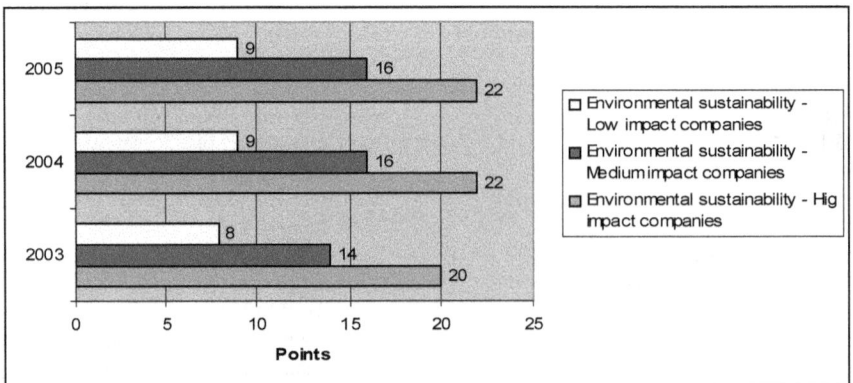

In addition, a score of at least a point, in relation to the core criteria of each pillar and category, was stipulated, starting from 2004. Core criteria are those the JSE considers fundamental, and not negotiable. The details regarding some specifications are as follows: at least five out of nine core criteria are in corporate governance; at least five out of eight core criteria relate to the environment; at least four out of six core criteria fall under the economic pillar; and at least seven out of ten core criteria are social (Johannesburg Stock Exchange 2004, 2005).

Sustainability Banking in Africa

This last section deliberates issues pertaining to sustainability banking in Africa. Financial institutions are increasingly being called on to safeguard against funding projects with negative environmental consequences. Much of the information presented here has been drawn from the 2004 landmark study by the African Institute of Corporate Citizenship (AICC), and ACCA's 2004 report entitled *Towards Transparency: Progress on Global Sustainability Reporting 2004* (ACCA 2004). The AICC survey included four countries (Botswana, Kenya, Nigeria and Senegal) and over fifty financial institutions (UNEP FI 2005). The AICC is a centre of excellence in corporate social responsibility 'committed to strengthening responsible growth and competitiveness in Africa through research, advocacy and network building' (African Institute of Corporate Citizenship 2004:83). The AICC describes sustainability in the banking sector as 'ensuring long term business success, while contributing towards economic and social development, a healthy environment and a stable society'. In this regard, sustainability has three broad components: people (socio-sphere), planet (enviro-sphere) and prosperity (econo-sphere). The need to address all the three spheres has been highlighted elsewhere.

The AICC has established that a number of sustainability banking practices were taking place in the continent, particularly from the case study countries. These included initiatives around pricing assets and exercising ownership, providing new finance, risk management, and savings and transactions.

Conclusion

This chapter has discussed issues pertaining to sustainability reporting as one of the emerging environmental management tools. Landmarks, including the Global Reporting Initiatives Sustainability Reporting Guidelines, were covered. An historical account of sustainability in business linked up company level sustainability reporting. The sustainability reporting nexus that harnesses concepts around corporate governance and sustainable development was also discussed, leading to further insights into the triple bottom line concept. A case revolving around the Johannesburg Stock Exchange Socially Responsible Investment Index was detailed. The final section gave an overview of sustainability banking in Africa.

Revision Questions
1. What is sustainability reporting?
2. Why should businesses account for environmental damage?
3. What are the fundamental provisions of the Global Reporting Initiative in terms of sustainability reporting?

Critical Thinking Questions
1. How has the concept of sustainability reporting assisted companies and governments in your country to holistically manage local and national environments?
2. Can the Johannesburg Stock Exchange Socially Responsible Investment Index model be replicated in your country? If not, what aspects could be amended to suit local conditions?
3. Has sustainability reporting compelled companies in your country to go beyond presenting a good public image in terms of good environmental stewardship?

References

ACCA, 2004, *Towards Transparency: Progress on Global Sustainability Reporting 2004*, London: Certified Accountants Education Trust for the Association of the Chartered Certified Accountants.

African Institute of Corporate Citizenship, 2004, *Sustainability Banking in Africa*, Johannesburg: African Institute of Corporate Citizenship.

Bennett, M. and James, P., 2004, 'The Green Bottom Line', in R. Starkey and R. Welford, eds., *The Earthscan Reader in Business & Sustainable Development*, London: Earthscan, pp. 126–57.

Elkington, J., 2004, 'The Triple Bottom Line for 21st-Century Business', in R. Starkey and R. Welford, eds., *The Earthscan Reader in Business & Sustainable Development*, London: Earthscan, pp. 20–44.

Ernest & Young, 2004, *Excellence in Corporate Reporting*, Johannesburg: Ernest & Young.

Esty, D. C., Levy, M.A., Srebotnjak, T., et al., 2005, *2005 Environmental Sustainability Index*, New Haven: Yale Centre for Environmental Law & Policy.

Esty, D. C., Levy, M.A., Srebotnjak T., et al., 2006, *Pilot 2006 Environmental Performance Index*, New Haven: Yale Centre for Environmental Law & Policy.

Fishbein, B.K., 1994, *Germany, Garbage and the Green Dot: Challenging the Throwaway Society*, New York: Inform.

Global Reporting Initiative, 2002, *Sustainability Reporting Guidelines*, Boston: Global Reporting Initiative.

Hart, S. L., 2004, 'Beyond Greening: Strategies for a Sustainable World', in R. Starkey and R. Welford, eds., *The Earthscan Reader in Business & Sustainable Development*, London: Earthscan, pp. 7–19.

IDSA, 1994, *Corporate Governance King I Report – 1994*, Parklands: Institute of Directors in South Africa.

IDSA, 2002, *Corporate Governance King II Report – 2002*, Parklands: Institute of Directors in South Africa.

International Atomic Energy Agency, 2005, *Energy Indicators for Sustainable Development: Guidelines and Methodologies*, Vienna: International Atomic Energy Agency.

Johannesburg Stock Exchange, 2005, *JSE SRI Index Background Selection Criteria*, Johannesburg: Johannesburg Stock Exchange.

Johannesburg Stock Exchange, 2004, *JSE SRI Index Background Selection Criteria*, Johannesburg: Johannesburg Stock Exchange.

Johannesburg Stock Exchange, 2003, *JSE SRI Index Background Selection Criteria*, Johannesburg: Johannesburg Stock Exchange.

Jonah, D. and Pienaar, K., 2004, *Introduction to Corporate Citizenship and Defending the Business Case: JSE Topical Seminars*, Pretoria: Centre for Corporate Citizenship.

Mahoney, M. and Potter, J.L., 2004, 'Integrating Health Impact Assessment into the Triple Bottom Line Concept', *Environmental Impact Assessment Review*, Vol. 24, pp. 151–60.

Pauli, G., 1997, 'Zero Waste: The Ultimate Goal of Cleaner Production', *Journal of Cleaner Production*, Vol. 5, pp. 109–113.

RSA, 1998, *National Environmental Management Act* (Act No. 107 of 1998). Pretoria: Government Printer.

Timberlake, L., 1992, 'Changing Business Attitudes', in J. Quarrie, ed., *Earth Summit '92: The United Nations Conference on Environment and Development Rio de Janeiro 1992*, London: The Regency Press.

UN, 1992, *Report of the United Nations Conference on Environment and Development*, New York: United Nations Secretariat.

UN, 2002, *World Summit on Sustainable Development Plan of Implementation*, New York: UN Secretariat.

UNEP FI, 2005, [pamphlet] *CEO Briefing: A document of the UNEP FI African Task Force. UNEP FI Sustainability Banking in Africa Report*, Geneva: UNEP FI.

United Nations Division for Sustainable Development, 1999, *Work Programme on Indicators of Sustainable Development of the Commission of Sustainable Development*, New York: United Nations Division for Sustainable Development.

United Nations Division for Sustainable Development, 2001, *Indicators for Sustainable Development: Guidelines and Methodologies*, New York: United Nations Division for Sustainable Development.

United Nations Division for Sustainable Development, 2004, *Assessment of Sustainability Indicators*, New York: United Nations Division for Sustainable Development.

Wixley, T. and Everingham, G., 2002, *Corporate Governance*, Cape Town: SiberInk.

Chapter 11

Promotion of Formal and Non-formal Environmental Education

Introduction

One important characteristic of environmental education is the employment of a wide variety of methods. This chapter considers some approaches and methods that could be employed in the implementation, monitoring and evaluation of formal or school-based, and non-formal or community-based, environmental education in any given setting. It begins with a delineation of the key paradigms that inform the formulation of environmental education approaches.

Some Paradigms of Environmental Education

Both formal and non-formal environmental education programmes and their various approaches will be discussed. It is therefore necessary to delineate some environmental education paradigms in order to consider those that inform the formulation of each of the approaches under discussion. Firstly, what is a paradigm? Lynch (2005) defines it as a lens through which we view the world. Accordingly, different lenses entail different assumptions about the nature of the world and the ways in which we should attempt to understand it. There are many different lenses that exist for viewing and understanding it. Interpretivism, eco-centricism, anthropocentricism and ecofeminism are some of the paradigms that have emerged to shape and influence approaches to environmental education at both formal and non-formal levels.

Interpretivism

Lynch (2005) provides a succinct description of the interpretivist paradigm. In this paradigm it is considered impossible to separate facts from values. Inherent subjectivity in any research conducted in relation to people or the social world is accepted. Since knowledge is socially constructed, rather than an independently existing reality, the notion of causality is defined differently. From the interpretivist perspective, causal relationships are simply another, possible explanation for certain aspects of the social world that is being researched. They are not taken to be universal laws that govern people and their actions. Rather than following the notion of causality as one variable preceding and causing another, interpretivism sees relationships as more complex and fluid, with directions of influence being mutual and shifting, rather than uni-directional and fixed. Relationships within the social world are external and independent of our attempts to understand them. Therefore, rather than seeking a 'true' match between our research observations and reality, the interpretivist paradigm understands reality as being constructed in and through our observations and pursuit of knowledge.

Eco-centricism

Eco-centricism (ecologically-centred) claims moral values and rights for ecological processes and systems, as an alternative to the biocentric view, which concentrates on the values and rights of individual organisms (Cunningham et al. 2003). In the eco-centric paradigm, the whole, for example, an ecological area or a species, is considered to be more important than its component parts, for example, an individual organism. A simple illustration given in support of this alternative view is that if you killed an individual organism you have only denied it a few months or years of life. But, if, on the other hand, you eliminated a whole landscape, you will have destroyed what has taken millions of years to create.

Anthropocentricism

This human-centred approach places the human above all other species, by virtue of the human being's intelligence, foresightedness and creativity (Cunningham et al. 2003). By extension, humans have dominion over all other species, and nature as a whole. This paradigm has its roots in the Bible. It is misinterpreted to mean that man has the right to exploit and treat nature as he pleases. It follows, therefore, that he can choose to protect only the species that serve his religious, cultural, scientific, economic and other interests. Of course, this has generally been applied. We may recall how often an insect, unfortunate enough to find itself on your arm, is crushed, even though you know very well that it can cause you no harm.

Eco-feminism

According to Cunningham et al. (2003), eco-feminism is a proposed alternative to the last two paradigms, which promote patriarchal systems of domination. Rather than concentrating on values, rights, obligations, ownerships and responsibilities, it encourages care, appropriate reciprocity and kinship. The rationale is that 'when people see themselves as related to others and to nature they will see life as bounty rather than scarcity, as cooperation rather than competition, and as a network of relationships rather than isolated egos' (Cunningham 2003:44).

According to Birkeland (1995), eco-feminism is a paradigm that is as broad as the patriarchal ones it seeks to supplant, positing nearly as many interpretations as there are eco-feminists. In a nutshell, however, it is the application of feminist theory to the ecological crisis. The basic notion is that social oppression and environmental exploitation are inextricably linked to fundamental social constructs that have co-evolved with patriarchal power relations. For example, in Western patriarchal thought, reality has been construed in terms of dualisms, such as culture/nature, reason/

emotion, subject/object, science/art, public/private, hard/soft, mind/body, and so on. These basic dualisms are gendered and hierarchical, one side of each opposition being associated with the feminine, and devalued in the culture. From an eco-feminist view, replacing this patriarchal, dominating culture with one of equality and mutual respect would yield greater benefits for all.

Formal Environmental Education Programmes

A formal or school-based environmental education programme considers results in the long- rather than the short-term. It is based on the assumption that the target population may grow up to become wise resource users, or informed policy- and decision-makers.

Where there is no syllabus for environmental education, the programme should start by involving some schoolteachers in the examination of existing curricula, with a view to identifying subject areas with environmental contents; that is subjects that have environmental issues. After this exercise, a syllabus or scheme of work should be prepared. At least one workshop should be organised to review the materials of the early draft, and an education authority should approve the final draft before it is put into use in schools. The approval of an education authority is to ensure that teachers are committed to the teaching of the subject. This brings us to another consideration: training workshops for teachers. It is generally bemoaned that environmental education is a new subject, which requires some level of teacher training in order to guarantee the effective teaching of the subject.

One fortunate thing about school-based environmental education programmes is the fact that it targets two controlled target audiences: the schoolteachers and the learners. These groups have a common interest, and a common position to defend their interests. The schoolteachers' interest is to ensure the success of the learners by doing effective teaching. The learners, on their part, struggle to succeed by studying hard. Many experts stress the need to carry out intensive environmental education programmes at primary school level. This is based on the fact that the primary school is a nurturing ground for children who are the future resource-users, as well as decision- and policy-makers. It is at this level that the teacher can inculcate proper environmental behaviour in children, who may then grow up to be wise resource-users. When placed in decision making positions, the children could make decisions that help conserve rather than destroy the environment. The primary school teacher, therefore, has a crucial role to play in improving the quality of the environment. She has the responsibility, and the ability, to mould children into citizens who are environmentally conscious, environmentally literate and environmentally responsible.

Approaches to formal environmental education

Although implementers of environmental education dispute the goal of this subject, since they often find themselves in situations that require different definitions, they tend to adopt one of the following approaches.

Separate subject approach

This means treating environmental education as a subject in its own right. The subject is included on the school timetable, and teachers prepare lessons and teach it as they do any other subject on the curriculum. The problem with this approach is that only few teachers are sufficiently well trained and informed to handle the subject. Another problem is that of curricula in many countries, that are already too crowded to permit the introduction of another subject.

Cross-curricular approach

This is known as the multi-disciplinary approach. It is the teaching of environmental education through all the subjects on the school curriculum. Existing school subjects are screened for topics of relevance to environmental education. Some Western experts advocate this approach. It satisfies those who see environmental education treated in a piecemeal and arbitrary manner in schools. Teaching the subject using this approach depends upon the interest of individual teachers and upon the imagination they bring into the interpretation of their subject syllabi (WWF 1988). If this becomes the popular approach, it should be noted that it involves a great deal of coordination with every subject teacher understanding their role, defining their input and ensuring that what each teacher complements the lessons of others, forming an overall coherent learning experience (Living Earth 1998).

Carrier subject approach

This is an abridged version of the cross-curricular approach. It uses one or two subjects to teach environmental education. It shares similar constraints with the cross-curricular approach.

Thematic approach

This is the teaching of environmental education through environmental themes, applied to the different school subjects. This approach can be quite demanding and needs highly skilled teachers, capable of developing, designing and managing a curricular content in an interdisciplinary fashion.

Project approach

This entails carrying out projects designed either to raise general environmental awareness or address identified environmental problems. Examples include the development of tree nurseries and the establishment of crop rotational farm plots, in the course of which relevant discussions relating to environmental care are carried out.

Methods for the promotion of formal environmental education

As has been noted elsewhere, environmental education ought to be all-embracing. This is reflected in the multiplicity of methods employed in the implementation of the subject, even at the level of the school. The following methods, discussed below, are by no means the only ones that can be used; the author has limited himself to those he has personally tested in schools.

Lesson presentations

As is the case with any other subject on the school curriculum, lesson notes are prepared on environmental topics and delivered to children using any of the approaches discussed earlier. Examples of lesson notes are included at the end of chapter four. A lesson presentation, or teaching, consists of a number of basic skills: communicating, explaining, questioning and organising (Farrant 1980). Each of these can be improved through constant practice. Demonstrations, illustrations and use of audio-visual materials are very important aspects of teaching, and there is increasing emphasis on student participation in the process.

Informal discussions

Teachers can discuss with the children in small groups or individually during breaks. This enables the children to develop the habit of free expression. It also helps build confidence in the children. Through informal discussions, the teacher can gain insights into children's personal values, attitudes, behaviour and potential. This is also a useful means of exchanging ideas and information on environmental issues.

Debates

It may be argued that, given the nature of most of our primary schools, this method may only be applicable at the secondary school level. But it should be noted that children of the senior primary classes can communicate freely and will have developed their own points of view by that stage. They often engage in arguments on their way to and from school and during breaks. This is a form of debate. Debates are an expression of opinion or position. Through debates, children learn to respect the views of others and to look at things from different perspectives. This is a good way of preparing them to grow into adults who can make sound judgements and informed decisions on environmental issues.

Drama

The students could be organised to perform short and simple environmental plays. Drama is an effective and entertaining way of putting across messages. It is the re-enactment of stories and presentation of situations, which makes the audience learn, as if by direct experience, and then draw conclusions by themselves. At the end of a drama performance, the audience should be asked questions in order to reinforce the meaning of the play, and prevent misinterpretations.

Games

Games, like drama, present messages in an entertaining and practical way. The rules of the game should be carefully explained, and instructions should be short and simple. As with drama, the children should be asked questions at the end of the game in order to reinforce the meaning, and avert misinterpretations.

Songs

Some groups of children could be organised to compose songs that convey specific environmental messages. These songs could be presented to the class during special lessons, to the entire school on special occasions, or to a wider audience during national and international day celebrations.

Nature clubs

Clubs provide a useful opportunity to not only develop environmental awareness in the children, but also to inspire them to take an active role in environmental conservation. The clubs could be organised to carry out activities such as clean-up campaigns, tree planting, and so on. Debates and drama could be organised by the clubs to raise environmental awareness in the entire community. Arrangements could also be made for the clubs to participate in national and international day celebrations, for example World Environment Day, during which simple but extremely important environmental messages displayed on placards and read aloud to the audience

Field visits

Field visits are an interesting means of bringing the schools in meaningful contact and dialogue with the environment. They provide opportunities for the pupils to purposefully explore and investigate some aspects of the environment, as well as appreciate its aesthetic values. They could be organised as follow-ups of lessons started in the classroom, or as bigger ventures involving members of school clubs, to explore themes of general interest.

Competition

Competitions can be a good way of encouraging or stimulating the expression of sensitivity towards the environment, and of developing skills that are useful for environmental conservation. They could eventually lead to spontaneous participation in environmental conservation activities. Competitions could be on environmental essay writing, environmental music, tree nursery establishment, tree planting, or even take the form of a quiz to test knowledge about the environment and environmental issues. Vefonge (2000) reports how a theatre competition on 'acting for the environment' was organised by the Mount Cameroon Project-Buea, to identify theatre troupes with the potentials to perform plays for the sensitisation of rural communities and the general public on environmental issues. Competitions could also be organised in the form of school projects. Projects are special activities carried out for particular purposes. They could be for fundraising, or to address or prevent common problems such as soil erosion. In the course of executing the projects, children acquire the knowledge and skills that would be useful to them in later life.

Non-formal environmental education programmes

Most conservation projects vouch for non-formal or community-based environmental education. This is due to the fact that these programmes directly targets the users of the natural resources to be conserved. It is also common for such programmes to be limited to conservation, hence the name 'conservation education' that some projects adopt. The programme aims to achieve short-term results, since most of them are carried out by short-term conservation projects, with no long-term commitments or agenda. This is an eco-centric (or wildlife-centred) approach, which may be misinterpreted as placing wildlife and their habitats before humans. Experience has shown that a programme that stresses this approach, more often than not, meets with total failure, as communities, sooner or later, start questioning whether wildlife is to be conserved at the expense of the human community. This brings us back to the necessity of a holistic approach, which should be a subtle blend of anthropocentricism (or people-centred-ness) and eco-centricism, in consonance with the definition of 'the best environmental policies' (Cunningham et al. 2003). This is shown in Figure 11.1. An anthropocentric approach includes elements that have an immediately direct benefit to the human species and are seen as such. This may include income generation or development activities, which could be employed as a strategy for encouraging the acceptance and participation of the target communities. The local people are more likely to participate if an activity is closely linked to the perceived benefits, and will continue to participate with even greater interest and enthusiasm if they started enjoying direct benefits from the effort.

In view of the above, it is advised that a community-based environmental education programme should include a reasonable balance of materials drawn from areas such as ecology, natural resource management, agriculture, health and community development. There should also be political and economic dimensions, but these should be handled with care, in order not to generate problems that may escalate beyond control.

The role of an environmental educator in community sensitisation, is more challenging than implementing a formal environmental education programme. For the educator to succeed with the communities, she must exercise tolerance, tactfulness and time-consciousness, since the programme targets a wide range of uncontrolled target audiences, each with an array of common and conflicting interests, and often with strong and unflinching positions to defend their interests. Providing motivation to move a community as a target audience from its present 'position' (which could be uncooperative or antagonistic) to one supporting and participating in the new effort seriously begs a definition. It is different from with school pupils, where the expected change is inevitable, since they know, from what and how they are taught, that they will be assessed on the basis of their demonstration of change in knowledge and attitude, and their ability to 'act' skilfully and responsibly towards the environment. Furthermore, they may be as yet too young to find themselves under the sort of economic and other pressures that could make such changes in behaviour difficult.

Figure 11.1: Best Environmental Policies

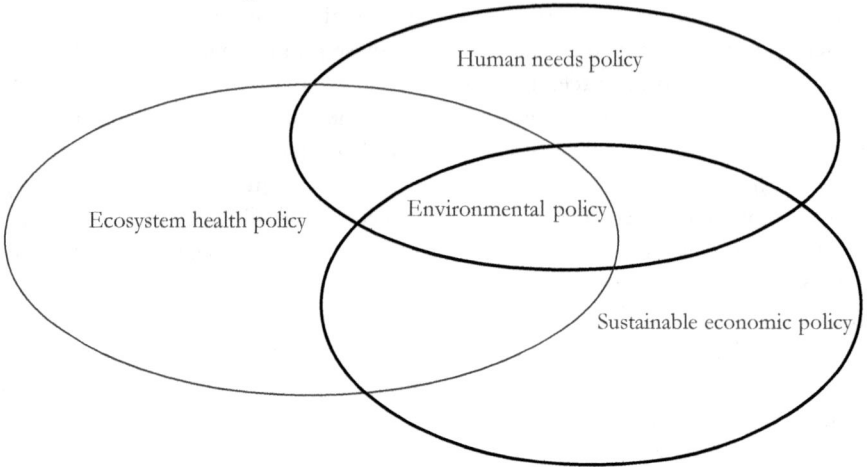

Source: Cunningham et al. 2003.

Approaches to Non-formal Environmental Education

As has been noted earlier, community environmental education programmes generally seek to address local environmental problems. This strategy of local problem solving presents an opportunity to face real environmental problems, since the environmental educator can prioritise the most pressing problems faced by the community and avoid undue emphasis on global problems, which are so distant from daily local reality that they provide little or no motivation for action (Laryargues 2000). The local problem solving strategy permits two types of approach: it may be applied as an activity end-in-itself, or as a theme generator.

Activity End-in-itself Approach

This approach aims to resolve a given environmental problem, after considering the associated ecological aspects. But, as Lary (2000) argues, it might foster a pragmatic type of education that provokes an attitude of problem solving, just to see the end results, rather than achieve the aim of stimulating a broader discussion of the problem. The power to mobilise resources and confront an environmental problem does not necessarily guarantee that the target participant understands the complex interactions of the ecological aspects of the problem with the political, economic, social and cultural aspects. The danger is that this will hinder reflection on the need to change society's cultural values. Put simply, there is no guarantee that once an environmental problem has been resolved by this approach, that he cause of the problem will not be repeated; since from this perspective, the ability to criticise the *status quo* has not been developed. This clearly supports Reigota's (1994) assertion that environmental education should prepare citizens to understand why something needs to be done, and, not merely, how to do it.

Theme Generator Approach

This approach evaluates the problem in a broader perspective. Instead of focusing only on ecological aspects, there is consideration of the political, economic, cultural and social factors inherent to the environmental problem. Although an understanding of how ecological systems work is necessary, if people are to understand the environment, a whole array of human factors, such as values, politics, culture, history and economics, must be explored if the current condition of the environment is to be properly understood (Martin 1990).

Methods for the Promotion of Formal Environmental Education

Meetings

Meetings are organised forums for discussions. They provide useful contacts with the target population. Once the first contact has been made, this should be maintained, regularly, as it helps to establish a solid foundation for mutual understanding, mutual trust, and good rapport, and to ensure a good working relationship with the target population.

Every meeting should have a purpose: it could be for planning, for sensitisation, for clarification of certain points, or simply for follow-up. Equally important is for every meeting to have something new to discuss, because discussing the same topic over and over again can be boring and may result in a drop in attendances during subsequent meetings. Discussions during meetings should, as much as possible, be participatory. Time should be given for participants to ask questions, or make comments, where necessary or indicated. Some questions asked can be thrown back at other participants to answer. This ensures active participation and reduces the tendency of some participants trying to place the educator as always in the answerable position.

Timing is a very important factor in scheduling meetings. The time of day, when it is most convenient for a wide range of participants to attend, should be taken into consideration. Organising meetings during periods when the communities are most busy, for example, when everyone is engaged in activities such as income generation or cultural events could be futile. The frequency of meetings during such periods should be reduced, and activities selected should serve to facilitate, or complement the current community activities. Furthermore, the purpose of the meeting should be clearly explained in a circular note, which should reach the target village at least a week in advance of the scheduled date. These circulars should be clearly addressed to the appropriate authorities in the community.

One-to-one Communication

There are some members of the community who are shy or afraid to express their views in the presence of others, especially in front of those with some level of influence and social status. One-to-one or individual communication provides a useful means of procuring information. But tact is necessary in order to avoid the impression of being intrusive. One-to-one discussions should take place in a friendly and fun atmosphere in order to ensure a free-flow of information. An educator who speaks in a condescending fashion, for instance, will normally not obtain adequate and reliable information, or feedback. Firstly, he will make the target person feel that he or she has nothing to contribute. This happens where the individual is shy and withdrawn. In the case of some bold and aggressive member of the community, the condescending attitude of the educator may generate a protracted argument that may gradually degenerate into a quarrel. Once this happens, the educator should be careful, because further similar incidents may seriously jeopardise the future of the programme. It is important to obtain feedback through the method of one-to-one meetings, because what may be seen as popular opinion may be the opinion only of a few influential members of the community. This may not be useful in achieving the intended goal of mass education, but influencing the behaviour of an individual at this level of interaction could have enduring effects.

Focus Group Discussions

Focus group discussions involve a group of people who share a common interest, for example farmers, to talk about a particular topic. This sort of discussion takes advantage of group dynamics, and allows respondents to be guided by a skilled moderator to explore key issues in greater depth (Margoluis and Salafsky 1998).

Workshops and Seminars

A workshop is an interactive kind of meeting devoid of preaching or lecturing. Each participant actively, or passively, contributes to and gets involved in the discussions and activities. It should be stressed that a workshop should be more or less activity-based, with the educator or any other competent person simply acting in the role of the facilitator. The discussions should be properly and carefully controlled and directed so that the workshop does not lose focus. The activities chosen should be meaningful and relevant to the workshop. This ensures that the intended and more revealing results are obtained.

A workshop could either be for training or for sensitisation. Whatever the purpose, there must be a well defined theme or topic. This should be communicated to the participants well in advance because they, too, have to make their own contributions. A good workshop should be able to expose participants to practical and logical ways of obtaining or understanding important facts and concepts, and also to developing basic skills necessary for the implementation of possible solutions to be identified

A seminar is a forum for delivering lectures or papers with a view to sensitising and educating the participants. An important role of the educator in this regard is in identifying good and competent guest speakers or lecturers. She also acts as a moderator, introducing the seminar topics and the speakers. She may or may not have to deliver a lecture herself, but has the significant role of overall coordination.

It is important for every presentation to be followed by a discussion session during which participants are free to express their own views, make comments or ask questions. Discussions should be controlled and directed. This should be done tactfully; questions or views that have no relevance to the topic should be discarded politely.

Communication and Information Resources

A successful education programme requires adequate and appropriate materials. These are useful in disseminating information and raising awareness. Some may serve to maintain communication links between the target populations, especially where the educator finds it difficult to make regular visits. Some commonly produced materials are newsletters, brochures, fact sheets, calendars, t-shirts, fez caps, and keyholders. All of these should be communicative, and the language and content should be appropriate to the level and nature of the target audiences. In the case of calendars, each month of the year could carry a specific environmental message in the form of short phrases translated into local languages, or artistic drawings and pictures. The latter would be most useful for illiterate members of the population.

Use of the Media

There is no doubt that the media have played a major role in raising public awareness of environmental issues (Boulton and Knight 1996). It may be advisable to collaborate with local newspapers, as well as with radio and television stations, for the release of articles and news items, respectively, on the environment. This, however, should be considered after the results of a survey, conducted to find out the percentage of the target population that makes use of these media forms, and about their frequency of use. This is to ensure that the effort is not wasted.

Mount Cameroon Project-Buea, in Cameroon, established what was known as Green Page in a newspaper, which served as a forum for debates and exchanges on environmental issues. Technical staff on the project opened up the debate with an introductory article that attracted reactions and contributions from the public, in the form of comics, cartoons, comments and poems. Those with knowledge of a given topic shared this with the readership. Those who were less informed about it simply asked questions to get information. The strategy was to gradually reduce the cost attached to the method, by making newspapers understand that environmental issues are interesting items for publication (Vefonge 2000). Radio and televisions stations have been known to independently report environmental issues. This can be used also to measure and monitor the increase in public interest about environmental concerns. For example, a survey could be conducted to find out how many items are being released on environmental issues monthly, and how many people listen to such programmes. This applies equally to the print media.

Use of the Arts

Some public awareness programmes make use of the arts, including street theatre, puppetry, fine arts, music and drama. Street theatre has been employed with tremendous success in India. It has been tried by the Cross River National Park environmental education programme in Nigeria, but with emphasis on the use of puppetry. Mount Kupe Project, Cameroon, is known to have employed fine arts. Paintings were made on the walls of an important building, depicting the local environment, where passers-by could see, admire, interpret and gain knowledge. A peace corps volunteer undertook a similar project by depicting various types of environmentally unfriendly activities on the wall of a nursery school building in Nguti, Cameroon. It was interesting to see how even passers-by, in a rush, stopped and spent time observing, especially if they were encountering the painting for the first time. The Wildlife Conservation Society in Cameroon once employed the talents of a local musician who composed and published conservation songs. This effort was emulated by Mount Cameroon (Vefonge 2000), and the Korup Projects; however their impact has not been assessed to reach any definitive conclusions as to its efficacy. However, given the role that music has played in political and racial campaigns, there is no doubt that the same results could be achieved in environmental education.

Of all the art forms, drama is by far the most widely used. Drama is a way of avoiding preaching that is either boring, or may breed suspicion, especially that of an obviously dogmatic nature. Drama provides a subtle and entertaining, but effective, way of communicating environmental messages. It is a way of re-enacting a story, real or imagined, not merely through words, but also with actions. This helps elevate the story to a concrete level, thus making the audience learn by direct experience, heightening their concentration (Figure 11.2).

Figure 11.2: Audience 'a' and 'c' Watching a Performance of The Sacred Forest by the Forest Pipers Drama Group of Nguti

(a)   (b)   (c)

These photographs were taken at Sumbe village in Southwest Cameroon.

Members of the audience may identify themselves with some of the characters and issues raised in the play. This can lead to a deeper understanding of the issues and messages conveyed. Inyang (2001) notes that communities have a strong motivation to attend drama, as most people see this as an opportunity for them to receive a rare form of entertainment; hence it is characterised by high attendances that include a wide range of target audiences. Similarly, it could have a positive impact on the community, since it provides an opportunity for a collective experience, which could lead to prompt, collective action.

An environmental play should not be too long, in order not to tire the audience and discourage them from attending subsequent performances. Also, the structure of the play should be simple and straightforward. There should be ample use of local idioms and metaphors; proverbs, songs, dance, and other traditional conventions (Inyang 2002) to give the play local colour and flavour. This facilitates communication and encourages audience participation. In terms of structure, the play should end in such a way that the 'possible worlds' encountered in the performance are carried back by the audience into the 'real' world, in ways which may influence subsequent action (Kershaw 1992). In other words, the play should permit and motivate the audience to 'complete' the performance, by implementing an action that they feel the performance has left undone (Inyang 2002). This is in violation of Hegel's and Aristotle's idea of a quiet somnolence at the end of the spectacle. It conforms to Brecht's vision that the end of the spectacle should be the beginning of action (Boal 1974).

The Use of Mobile Education Units

Mobile education units, though not very common, due to the costs involved, are an innovative method of taking environmental messages to the local communities and the general public. A vehicle is equipped with film, slides and/or video projection equipment, as well as with other educational and informational materials that could be carried or displayed on the vehicle for people to collect or view. Other equipment of the unit includes a small generator to provide electricity for the operation of the film, slides and video appliances.

Involvement of Local Communities

Commonly owned resources, in open access regimes such as forests, lakes and rivers, are difficult to manage, as each individual strives to maximise their profit from the common pool, with the concomitant destruction or degradation of the resource base (Hardin 1968). In such a situation the use of enforcement as a regulatory mechanism is often proposed (Lewis 1996). But using this strategy in the absence of other incentives would almost certainly result in 'passive' rather than 'active' conservation (Ferraro and Kramer 1995). Passive conservation means that the local communities might themselves desist from carrying out the illegal activities, for fear of being prosecuted. But they would not report cases of, let alone take action against, any such activities. This leads to the conclusion that many environmental problems cannot be solved without the participation of the local communities. Participatory approaches have the incredible advantage that they give the environmental educator a better understanding of local values, knowledge and experience. They win community backing for project objectives, and communities can help with the implementation of activities and resolution of conflicts over resource-use (World Bank 1992).

One important approach to community involvement is facilitating the formation of clubs and committees in village communities, with functions ranging from community health and sanitation to natural resource management. Inyang (2003) warns that in order for the communities to assume full responsibility for all major decisions about group formation, the allocation of tasks and implementation of activities should be based on their own careful reasoning. Furthermore, it is advisable for these committees to work under traditional councils, and in close collaboration with other traditional institutions, as well as with relevant local government institutions. This arrangement guarantees that the committees are recognised, empowered and supported in every way possible, especially if and when it is necessary for them to implement law-enforcement, or carry out activities such as tree planting and other environmental campaigns.

When members join a club, or any other type of organization, they have given up some personal freedom of action, as part of the price of membership. They are obliged to adhere to established norms, usually defined by internal rules and regulations. These, together with other tools and mechanisms, facilitate socialisation, which

Buchanan and Huczynski (1991) define as a process through which an individual's pattern of behaviour and their values, attitudes and motives are shaped to conform with those considered desirable in a particular organisation or society. Exposure to workshops and seminars is a good way of equipping the club or committee members with the necessary techniques and skills. The environmental educator might be required to assist in the drawing up of programmes. This ensures that the programmes are drawn to reflect the environmental situation, and that the activities are chosen take into account the needs of the target communities, as well as the objectives of the overall programme. Only this approach guarantees solutions with long-term results.

Membership of the clubs and committees should be voluntary. But it would not be a wrong approach to do some campaigning, by explaining the functions and importance of the clubs and committees. No one will join an organisation whose functions are not understood. There should be a registration fee, however small. This serves as a sort of pledge for active membership and commitment. It is recommended that membership cards be made, and every member given one upon registration, if possible. It may also be advisable to register the clubs and committees with the relevant government institutions, in order to give them legal backing. Once a club or committee has attained an appreciable level of development, it should be allowed to operate independently. This guarantees continuity. But engendering this independence requires maximum care; hence the need to monitor the progress of the club or committee for a reasonable period of time, after the first signs of its ability to operate independently. This is the stage in the club or committee's development, when some inherent problems will be brought to light, as we shall see below.

Unlike economic interest groups whose efforts directly benefit group members – since they offer private or individual goods that must be paid for by non-members, the efforts of non-economic groups such as environmental clubs and committees discussed above largely benefit everyone in the community. This is so because these groups more or less offer public or collective goods, giving rise to what is referred to as the free-rider problem (Patterson 1993; Baden 1998; O'Toole 1998). This describes a situation where some group members become inactive since they, like non-members, can benefit from the goods provided, without paying for the costs. By withholding their contributions to the group's efforts, these free-riding members are likely to be ahead of the other members since they can invest their contributions of time and effort in some other personal business (Baden 1998). In order to deal with the free-rider problem, there should be powerful incentives for active participation. Such incentives include opportunities for active members to enjoy the privileges of securing jobs, or being appointed to some important positions in the community (Patterson 1993).

Monitoring and Evaluation of Environmental Education Programmes

Monitoring and evaluation is an important aspect of many programmes, albeit often neglected or inadequately developed. It is like setting an alarm warning system to alert when things are going wrong or moving in the wrong direction. In order for a programme to succeed, a monitoring and evaluation system should be incorporated into it at the planning stage. At this stage, the goal and objectives of environmental education and the indicators of success should be clearly defined. All the activities of the programme should together achieve the five environmental education objectives, already discussed. Therefore, it is important to know which activities serve to achieve which objectives. In many cases, environmental education programmes have failed to be convincing to donors and managers of projects and institutions, largely because of the absence of well developed and well defined monitoring and evaluation programmes.

Policy-making stakeholders tend to place much emphasis on tangible results from environmental education initiatives, as justification for their usefulness, without necessarily considering how long it might take for such results to materialise. But, as Encalada (1992) warns, the effectiveness of environmental education cannot be only perceived in relation to direct changes in the environment. It must be observed in the way individuals and institutions are applying methods, procedures, and mobilising actions, known to bring changes in the environment.

It is important to note that although monitoring and evaluation are two sides of the same coin, they each serve slightly different purposes. Monitoring is a continuous and systematic process of checking whether activities are carried out as planned. It is a regular or periodic collection, by the staff, of information that can be used to make incremental adjustments to the programme or project (Stone 1997). Evaluation, on the other hand, is a process through which information is collected and analysed with the purpose of judging how well programmes have achieved, or are achieving, their objectives. It occurs usually in the middle, or at the end, of a project. Baseline information is necessary, as this provides the basis for determining changes. It is conducted either by the staff, or people outside the organization. It is based, to a large extent, on the results of monitoring. Its aim is to make major changes to the programme. The following are some of the methods commonly used in carrying out monitoring and evaluation of environmental education programmes.

Literature Review

Depending on the availability of relevant literature on the area, conducting a literature review is the first important step in information gathering. Information from the literature may give a fair picture of the biophysical situation of the area: soil and vegetation types, the drainage system, the wildlife resources, etc.; the socio-political structure and nature, including the important traditional institutions and their functions; and the predominant economic activities, such as farming, hunting and harvesting of timber and non-timbers forest products. In a school setting, this may consist of collating records such as teachers' lesson notes, and students' progress reports. It is also of absolute necessity to find out the types of questions that the students are asked, or about the activities in which they are engaged, for the assessments because numerical scores alone are not sufficient to give the sort of picture needed to draw informed conclusions.

Information gathered from literature points to what is expected, and not necessarily what prevails, at the given moment, because, depending on the age of the literature, events and time will have changed the picture. Conclusions cannot be drawn based entirely on information obtained from literature. It is expedient to cross-check this information with firsthand experiences, before drawing definitive conclusions. Another important piece of advice is that no matter how much you will have learnt about the area from literature, you should endeavour to approach your target community with an open mind, in order to have the opportunity to gather information that reflects the *status quo*.

Observations

Observation can be defined as a personal assessment of a situation, activity or behaviour in order to obtain basic information. This may take as long as a couple of years, and as little as a few minutes, depending on the type of information being gathered. There are two types of observation: direct observation and participant observation. Direct observation refers to independent assessment carried out without probing anyone for facts. This is usually done without the knowledge of the person or thing being observed, and usually happens by chance. It provides suitable means of obtaining information that is reliable, because of its potential to uncover the real situation, as those observed are going on with their activities naturally.

Participant observation depends on the collaboration of others to obtain the necessary information. The observer arranges to accompany the person to be observed to their area of activity, and observes how the activity is carried out. Usually the person under observation is informed about it, in the hope that this might prevent any modification of behaviour or activity, due to suspicion. The observer poses a series of probing questions in order to obtain additional, relevant information about the activity. Again, this should be done tactfully in order not to raise suspicion. However skillfully the observation is carried out, there is always the danger that, being aware of what is happening, the person being observed may still alter his or her behaviour: to appear more competent, more enthusiastic, more diligent, just to help the observer. As a result of these 'reactive effects', observations may be artificial, or unreal, and, therefore, a false reflection of the true nature and behaviour of the observed (Buchanan and Huczynski 1991).

In carrying out any of the two forms of observation, it is advisable that the observer is open-minded and unbiased. There should be no application of preconceived ideas, when drawing final conclusions. Furthermore, in planning the exercise, consideration should be given to the seasonality of activities in the communities, which form part of the monitoring units. This is to ensure that it is the right moment to conduct the observations. Seasonal events and activities calendars, which can be prepared during Participatory Rural Appraisal (PRA) sessions, with the communities, are useful tools in guiding planning decisions in this regard.

Questionnaires

A questionnaire is similar to an interview, since it is concerned with getting feedback from target audiences. The only difference is that questions are written or typed on sheets of paper, which are then distributed to the members of the target audience for whom the questionnaire is intended. It might be advantageous to bring the members together in a group, and explain the *raison d'être* of the questionnaire, before distributing the papers to them. Where there are illiterate members, some trustworthy literate members could be asked to assist. But care should be taken to ensure that the responses are sincerely those of the respondents. Some questionnaires are prepared and administered in a special way, with a view to assessing environmental attitudes, and are generally referred to as attitude surveys. For a questionnaire to be administered effectively, the questions should be simple and straightforward, requiring choices from a list of options, or for short answered to be provided, and should have variety in type and structure. The sequencing of the questions is also of vital importance, and should be well thought through. This is to ensure that, as much as possible, the respondent is encouraged to give honest and clear answers, which do not present difficulties in data analysis.

Semi-structured Interviews

This is an informal type of interview, conducted to gather useful information on a given subject. It takes place in a cordial and fun atmosphere, and allows for the free expression of the interviewee. Starting with an ice-breaker, to bridge the gap between the interviewer and the interviewee, could help create a conducive and enabling atmosphere.

The interviewer may make use of a set of open-ended questions, referred to as an interview guide. But this should not be used in a manner that destroys the purpose of the interview. An interview should be allowed to progress in a natural and conversational fashion. Some people are often tempted to record interviews on tape; this can be dangerous in certain situations. If the atmosphere becomes too formal, the interviewee may become nervous, inhibiting reasoning and impeding effective communication. It may also raise the suspicions of the interviewee, thereby preventing him or her from giving honest answers, or accurate facts. Information collected from interviews is usually more in-depth and richer, but more difficult to analyse, than information collected from questionnaires.

Written Tests

These mostly apply to the formal environmental education programme. A set of environmental questions can be developed, and the pupils are asked to provide answers in an examination setting and atmosphere. This is mostly used where environmental education is a free-standing subject on the school timetable, and is one of the subjects in which the students are being officially assessed. However, even where this is not yet the case, conservation or other projects in the area that help with the promotion of the subject could encourage teachers of participating schools to conduct periodic written tests. Depending on the techniques used in setting the questions, tests can provide a useful means of measuring not only pupils' knowledge, but also their attitudes and behaviour towards environmental concerns.

As a general principle, tests may be administered at the beginning of the school year (pre-test) to obtain baseline information about the target classes, and at the end (post-test) to measure how much the participants have learned. However, in-between, periodic tests are necessary for the purpose of monitoring progress.

Practical Tests

These are applicable to formal education programmes. They are exercises designed to test the student participants' new knowledge and skills in a real, or quasi real, context. They can take place during field visits, or through simulations in the classrooms or school compounds. The examinees may be asked to perform tasks ranging from, for example, enumeration of some of the natural habitats, and identification of the associated environmental problems, to analysing the problems, and suggesting and describing practical activities to address the identified problems. The simplicity or complexity of the tasks involved in the practical tests depends on the level of the examinees. As with written tests, practical tests could be used also to assess environmental attitudes and behaviours. If conducted in atmospheres free of the, often inhibiting, 'examination fever', the students will be able and motivated to express and conduct themselves in ways that portray their level of passion for the environment. However, assessing attitudes using this means is rather subjective for any conclusions to be drawn; and needs to be supplemented with the results of recent attitude surveys.

Revision Questions
1. Differentiate between a non-formal and formal environmental programme, and discuss the approaches and methods of each.
2. What tools can be used in the monitoring and evaluation of an environmental education programme?

Critical Thinking Questions
1. What measures can you propose to bring about the integration of environmental education into the school curriculum?
2. Discuss how the following environmental education approaches are influenced either by eco-centric, anthropocentric and ecofeminist paradigms:
   a) project approach, b) theme generator approach, activity-in-itself approach, and d) cross-curricular approach.

References

Baden, J.A., 1998, 'A New Primer for the Management of Common-pool Resources and Public Goods, in J.A. Baden and D.S. Nooman, eds., *Managing the Commons*, Bloomington and Indianapolis: Indiana University Press.

Boal, A., 1974, *Theatre of the Oppressed*, London: Pluto Press.

Boulton, M.N. and Knight, D., 1996, 'Conservation Education', in Spellberg, I.F., ed., *Conservation Biology*, London: Longman, pp. 69–79.

Buchanan, D. and Huczunski, A., 1991, *Organisational Behaviour: An Introductory Text*, Essex: Pentice Hall.

Birkeland, J., 1995, 'Disengendering ecofeminism', *Trumpeter* Vol. 12, No. 4.

Cunningham, W.P., Saigo, B.W. and Cunningham, M.A., 2003, *Environmental Science: A Global Concern*, New York: McGraw-Hill.

Encalada, M.A., 1992, 'EDUNAT: The Environmental Education and Awareness Raising Experience of Fundacion Natura in Ecuador', in Schneider, H., Vinke, J. and Weekes-Vagliani, W., eds., *Environmental Education: An Approach to Sustainable Development*, Paris: Organisation for Economic Co-operation and Development, pp. 215–26.

Farrant, J.S., 1980, *Principles and Practice of Education*, Essex: Longman.

Ferraro, J. and Kramer, A., 1995, *A Framework for Affective Household Behavior to Promote Biodiversity Conservation*, Arlington: Environmental and Natural Resources Policy and Training Project.

Hardin, G. 1968, 'The Tragedy of the Commons', *Science*, Vol. 162, pp. 1243–48.

Holden, P.J. and Inyang, E., 2001, *Drama for the Sensitisation and Mobilisation of Local Communities for Conservation: Lessons from the Banyang-Mbo Wildlife Sanctuary Project*, an unpublished paper prepared for the Banyang-Mbo Wildlife Sanctuary Project.

Inyang, 2002, Community Drama: An Alternative to the Troupe-Dependent Environmental Drama at the Banyang-Mbo Wildlife Sanctuary in Southwest Cameroon, an unpublished paper presented at the University of Strathclyde.

Kershaw, B., 1992, *The Politics of Performance: Radical Theatre as Cultural Intervention*, London: Routledge.

Layrargues, P.P., 2000, 'Solving Local Environmental Problems in Environmental Education: A Brazilian Case Study', *Environmental Education Research*, Vol. 6, No. 2, pp. 166–78.

Lewis, C., 1996, *Managing Conflicts in Protected Areas*, IUCN Biodiversity Programme.

Living Earth Cameroon, 1998, *A Teachers' Foundation Unit on Environmental Education*, Living Earth Foundation.

Lynch, B.K., 2005, *The Paradigm Debate*, <www.iltaonline.com/newsletter>

Margoluis, R. and Salafsky, N., 1998, *Measures of Success: Designing, Managing, and Monitoring Conservation and Development Projects*, Washington DC: Island Press.

Martin, P., 1990, *First Steps to Sustainability: The School Curriculum and the Environment*, Surrey: WWF.

O'Toole, R., 1998, 'The Tragedy of the Scenic Commons', in Baden, J.A. and Nooman, D.S. eds., *Managing the Commons*, Bloomington and Indianapolis: Indiana University Press.

Patterson, T.E., 1993, *The American Democracy*, New York: McGraw-Hill.

Reigota, M., 1994, *O que é educação ambiental,* Sao Paulo: Brasiliense.

Stone, R., 1997, *What's Your Role? Training for Organisational Impact. A Guide for Training Officers in Protected Area Management, Africa Biodiversity Series,* No. 5, Washington DC: Biodiversity Support Program.

The World Bank, 1992, *The World Development Report 1992,* Washington DC: Oxford University Press.

Vefonge, N., 2000, Environmental Communication in Mount Cameroon Project–Buea and the Need For Collaboration with Other Conservation Projects in the South West Province, a paper [unpublished] presented at the South West conservation projects environmental educators meeting held in Kumba, Cameroon.

WWF, 1988, A Common Purpose: Environmental Education in the School Curriculum, Surrey: WWF.

PART III

EMERGING ENVIRONMENTAL ISSUES

Chapter 12

Conclusion: Emerging Issues
and the Way Forward

We may recap that there still remain highly sensitive discourses on how best the African continent can speak in one voice in order to address some of the key environmental problems. An example is the importation of hazardous waste, with AU members having ratified the rival Basel and Bamako conventions. Africa has embraced the concept of sustainable development as a development paradigm for this century. Sustainability, or sustainable development, embraces politics, the environment, economics, culture and society. However, there is need to view sustainable development as an open question requiring continuous deliberations and debate. To this end, the issues that need to be addressed concern ways in which the AU should set up institutions and funds aimed at addressing continued environmental decay.

The issue of climate change appears remote, compared with such immediate problems as poverty, disease and economic stagnation. Development planners are often unsure whether, and how, to mainstream climate change considerations into development objectives (Agrawala 2005). It should be emphasised that the objective of the framework convention on climate change convention stresses the need to ensure that food production is not threatened, and to enable economic development to proceed in a sustainable manner (UN 2005).

The ratification of the UN framework convention on climate change was motivated by strong concerns that human activities are substantially increasing the atmospheric concentrations of greenhouse gases, and that this will result, on average, in additional warming of the earth's surface and atmosphere, with potential adverse effects on natural ecosystems and humankind (UN 2005). In spite of evidence in support of climate change, this issue has remained one of the most controversial in international negotiations since the 1980s (Shimada 2004). While assessments of past and present emission patterns strongly influence the debate over international climate policy, the central challenge is to limit future emissions (Baumert, Pershing et al. 2004). The Clean Development Mechanism (CDM) gives developing countries an opportunity to get involved in the implementation of the Kyoto Protocol, with the aim of meeting the emission reduction targets of the polluting Annex I countries, and the development needs of the developing countries. However, many methodological and other issues regarding the implementation of CDM activities, including the role of afforestation and reforestation projects, remain unresolved.

Given the prevailing stable investment climate in Africa, particularly in Morocco, Mozambique and South Africa, CDM projects will remain an attractive venture. However, due to various threats associated with the CDM – risks, uncertainty, ignorance in some sections of community and indeterminacy – Rio's precautionary principle should be applied in all dealings with the investing Annex 1 countries. The following suggestions are therefore set out: There is a need to enact firm and flexible stakeholder-driven CDM regulations at national, provincial and local government levels. The continent still has varying views regarding the best way forward concerning the CDM. Inter-ministerial and departmental coordination at country level remains chaotic and very weak, with regard to CDM project implementation.

The use of CDM project quota system is needed to safeguard biased investment in sectors considered 'low hanging fruit', of easy carbon credit picking, such as landfill gas and hydropower.

Capacity building and sustained awareness raising programmes are required at all levels of government, including traditional and community leadership. More local financial resources must be mobilised. For countries like South Africa this could be done through the King II Report and the JSE Socially Responsible Investment Index, whereby government and other local players earn 'carbon credits' for the benefit of the environment and future generations even outside the CDM. Dialogue is needed to continuously engage with the CDM and raise questions.

Although they cover only between 6 and 7 per cent of the earth's landmass, rainforests provide a habitat for about 50 per cent of all known species (World Bank 1992). The tropical rain forests are disappearing at rates that threaten the economic and ecological functions they provide. Shifting cultivation is considered to be the major cause of deforestation in Africa (Bundestag 1990), followed by logging, to meet the extremely high rates of timber consumption by the industrialised nations (Struhsaker 1998; Durning 1992). Tropical countries often struggle under massive debt loads, which drain their viability and encourage them to liquidate their forest capital more quickly to raise foreign exchange (WRI 1992). Additionally, logging offers an easy means of providing access roads to the rural areas, and easily wins rural community and national support.

Wildlife-human conflict, characterised by crop raiding by wildlife, is becoming an increasingly serious problem in Africa. This conflict can be attributed in large part to rapid human population growth, and poor land-use management strategies, which impose increasing demand on land. This demand results in an ever-increasing encroachment on wildlife habitats by agricultural activities and the development of human settlements. This phenomenon is explained by Hunter (1996), who postulates that the geometry of natural habitat fragmentation, induced by agriculture, indicates that as the wildlife range contracts in the face of human expansion, the interface of potential wildlife-human contact increases. Inyang (2002) observes that even hunting, which communities often employ as a strategy for addressing the problem of crop raiding by wildlife, instead helps to perpetuate it, as the targeted species are forced to disperse wildly, and find refuge in farmlands, due to the increased disturbance in the forest. Furthermore, the land question remains unresolved. Countries like Namibia, South Africa and Zimbabwe are still battling with how best to resolve the land question.

Over-exploitation is characteristic of open-access regimes. It results in economic warfare, which, at the micro-level, forces individuals to increase their rate of exploitation in order to maximise profits (Hardin, 1998). At the macro-level, economic warfare is orchestrated by 1) the struggle by the developed countries to maintain their position at the top rungs of the economic ladder in order to continue to enjoy their dignity as super powers; 2) the struggle by developing countries to also reach the top rungs of the global economic ladder in order to gain recognition as super powers; and 3) the struggle by the undeveloped countries to get out of poverty in order to seek ways of developing and liberating themselves from an aid-dependent mentality.

Natural resource management (NRM) has become an important preoccupation of NGOs and governments around the globe. The role of NGOs in encouraging the devolution of power to rural communities by reluctant governments, and in building the capacities of the communities, in NRM is enormous. But despite all their efforts, problems still abound in the management of natural resources in many parts of Africa.

It is increasingly acknowledged that the success of natural resource management (NRM) depends on the involvement and active participation of the communities that traditionally have rights of access to, and use of, the resources to satisfy their basic necessities; hence the term community-based natural resource management. However, as Inyang (2005) warns, community participation is closely linked to perceived or tangible short-term and long-term benefits, not only to the entire community, but also to actively participating community members.

Gender issues are gradually, but steadily, taking centre stage in natural resource management, especially because the livelihoods of rural women and men are intimately linked to natural resources. Whereas men are more prominent in activities such as hunting and fishing, women are the primary exploiters of non-timber forest products such as *Invingia gabonensis*, *Gnetum africanum* and *Ricinodendron heudelotin*, for both domestic consumption and income generation. Women are also engaged in trade in these and other non-timber forest products, like game meat which is also sold by some women as pepper soup. The interactions of women with the environment are in no way less important than men's interactions, in terms, particularly, of their impact on both the environment and the human community. Therefore, the exclusion of women from environmental and natural resource management can have negative impacts at both household and community levels (Commonwealth Secretariat 1996).

The African continent has made significant strides towards addressing sustainability principles through the application of EIA as one of the key decision making tools for approving development projects. Significant gains have been recorded in terms of establishing legislation, specifically addressing EIA requirements, in various countries. However, the following still remain slippery issues as regards the fine tuning of EIA procedures in the continent (El-Fadl and El-Fadel 2004; Kakonge 1993, 1994, 1998; SAIEA 2004).

Public Participation

The public still needs to be made aware of their environmental rights; governments need to open up when it comes to debating issues of good governance, so as to encourage participation in dealing with environmental matters in the EIA process.

Local Government Blackouts

Many local authorities are not directly responsible for EIAs. Yet most developments are implemented within their jurisdiction. In addition, local authorities have traditionally controlled development through various regional, town and country planning acts, which by their nature comprise considerable elements of EIA. In this regard, we recommend that efforts be made towards decentralising EIA permitting authority to local authority, so that harmonisation might be worked out between town planning and EIA laws; lastly, to cut red tap.

Harmonisation of EIA Legislation

At national, sub-regional and AU levels: the key challenge for Africa governments to harmonise EIA legislation at all levels still remains, culminating into an African Union 'mother' EIA legislative framework. EIA laws at national levels are still highly sectoral. Yet sub-continental frameworks, in eastern, central, northern, southern and West Africa, can be put in place, and can eventually feed into one EIA legislative framework within NEPAD, or at AU level, as an EIA convention.

Selective Sectoral Application

Most EIAs are applied to specific development sectors and even, specific projects within the sectors that are perceived to have severe negative impacts (Tarr 2003). Sectors traditionally exposed to EIA in Africa include mining, petroleum and gas; as well as agriculture, but mainly limited to dams. Agricultural policies seldom receive EIA attention, yet these have potential to harm the environment. Zimbabwe's 2000 Fast Track Land Reform Programme is an example of an environmentally harmful agricultural policy. Fisheries and tourism projects receive limited attention, likewise.

Expertise in EIA

As of June 2002, the whole of SADC had only eighty professionals managing EIA institutions (SAIEA 2004). Most tertiary institutions do not have courses addressing environmental management in general, and EIA specifically. Furthermore, government departments have experienced severe 'brain drain' on a national, regional and international scale. Experienced EIA professionals often switch jobs to join better

paying private and NGO sectors. Therefore, more effort should be made to encourage the establishment of courses in this area. Resource pooling can also assist in
utilising the available limited EIA expertise through initiatives, which seek to form
coalitions between governments, NGOs, private sector, universities and other research institutions. As in the health profession, issues pertaining to the environment
and EIA must also be prioritised.

Enforcement

Most legal documents do not stipulate clear monitoring and auditing procedures for
EIAs, nor prescribe resultant penalties to offenders thereof, additional to the lack
of monitoring and auditing of EIA (Ahmad and Wood 2002). This area needs
urgent attention. Regular monitoring is necessary (Tarr 2003) to ensure that developers implement agreed management plans. South Africa (DEAT 2004) is one of
the countries to have taken issues of compliance and enforcement seriously, establishing for example Environmental Courts in 2004.

Under-resourced EIA Institutions

Many in positions of authority, in politics and business, still consider EIA to be
another unnecessary hurdle that delays development, job creation, and ultimately
poverty eradication in the continent. There is therefore a need for continued lobbying, particularly from peers who realise the benefits of engaging in EIA.

Sectoral Orientation to EIA

EIAs are still undertaken and driven from sectoral point of view. Hence many
government ministries and departments consider EIA to be solely the responsibility
of the ministries responsible for environment and tourism. A cross-cutting paradigm is therefore being advocated in this book.

Logistics and Team Management During EIA Preparation

Drawing from a large-scale EIA for the proposed Dune Mining at St. Lucia in South
Africa, Weaver et al. (1996) note that EIA teams need to complement each other,
not only technically, but also in their purpose. Expectations and approach to the EIA
should also be mutually understood, and all members should be mutually accountable for their joint efforts. Logistical issues, particularly around public participation,
are usually seen as delaying the process.

Recognition of Potential Sub-regional EIA Promotion Initiatives

Governments should recognise and resource sub-regional EIA initiatives. One good
case example is the initiative by the Southern African Institute of Environmental
Assessment (SAIEA), an indigenous NGO based in Namibia. SAIEA is dedicated to
promoting EIA as a tool to achieve sustainable development and eradicate poverty
in southern Africa. Through partnerships, SAIEA has been supporting government,

development agencies, other NGOs and the private sector in the field of EIA. Some of the support mechanisms offered by SAIEA include: developing terms of reference for EIAs, independent reviewing, monitoring implementation, training (including hosting student attachments or interns), and research and assisting with EIA legislation reform and formulation.

Environmental education is recognised not only as one of the instruments in the fight against environmental problems, but also as a precondition for introducing rational management of environmental resources needed for human survival (Touré (1993). Despite this recognition, the subject has hardly been given the scope and dimension it deserves in any single situation, due largely to disputes that exist about its goal. Schneider (1993) attributes this to the fact that in any given local or national situation, environmental concerns of individuals, groups of people, and even institutions, tend to be limited, rather than all embracing. For instance there is a tendency in site-specific situations to limit environmental education to areas such as agro-forestry, conservation education and hygiene. This is contrary to Martin's (1990) guiding principles that advocate for environmental education to consider the environment in its totality, thus, as natural and man-made, ecological, political, economic, technological, social, legislative, cultural and aesthetic. Environmental education should enable people to develop an understanding of the ecological processes that govern life on earth, and the geo-morphic and climatic patterns that influence living things and human activities; appreciation of the social, economic and cultural influences that determine human values, perceptions and behaviour; and awareness of an individual's personal relationship with the environment, as a consumer, producer and a sentient member of society (WWF 1988).

While awareness raising engenders consciousness of the existence of environmental problems and the causes, sensitisation leads to the cultivation of positive environmental attitudes, which can translate into positive environmental behaviour. Cultivating positive environmental behaviour depends on the level of sensitisation of the individual. This increases with age, not merely in biological terms, but with respect to an accumulated attachment to particular ideas and values; as well as through the powerful influences of the family and peer groups, rigorous norms and belief systems of the community, patterned indoctrination through programmes of the mass media, schools and similar institutions, or emerging economic and related opportunities (Patterson 1993).

Environmental education has developed significantly in southern Africa, spearheaded by the formation of the Environmental Education Association of Southern Africa (EEASA) in 1982. Since then, the grouping has grown from strength to strength, leading to the hosting of regular annual conferences throughout the region. The EEASA also hosts the *Journal of Environmental Education*, the only such journal from the African continent published regularly on an annual basis. Lastly, the concept of sustainability reporting remains one ot the new environmental management tools with which Africa is still grappling. This concept has been significantly developed in South Africa, resulting in the launch, in 2004, of the Johannesburg

Stock Exchange Socially Responsible Investment Index. It remains to be seen whether, beyond its public relations benefit, this new concept will propel companies to truly address environmental concerns.

References

Agrawala, S., 2005, 'Putting Climate Change in the Development Mainstream: Introduction and Framework', in Agrawala, A., *Bridge Over Troubled Waters: Linking Climate Change and Development*, Paris: OECD, pp. 85–132.

Agrawala, S., Gigli, S. and Raksakulthai, V., et al., 2005, 'Climate Change and Natural Resource Management: Key Themes from Case Studies, in Agrawala, S., *Bridge Over Troubled Waters: Linking Climate Change and Development*, Paris: OECD, pp. 85–132.

Ahmad, B. and Wood, C., 2002, 'A Comparative Evaluation of the EIA Systems in Egypt, Turkey and Tunisia', *Environmental Impact Assessment Review*, Vol. 22, pp. 213–234.

Baumert, K., Pershing, J., et al., 2004, *Climate Data: Insights and Observations*, Washington DC: Pew Centre on Climate Change.

Commonwealth Secretariat, 1996, Women and Natural Resource Management: An Overview of a Pan-commonwealth Training Manual, London: Commonwealth Secretariat.

DEAT, 2004, *Environmental Impact Assessment Regulations*, Pretoria: Government Printer.

Durning, A., 1992, *How Much is Enough? The Consumer Society and the Future of the Earth*, New York: W.W. Norton and Company.

El-Fadl, K. and El-Fadel, M., 2004, 'Comparative Assessment of EIA Systems in MEAN Countries: Challenges and Prospects', *Environmental Impact Assessment Review*, Vol. 24, pp. 553–93.

Hardin, G., 1968, 'The Tragedy of the Commons', *Science*, Vol. 162, pp. 1243–48.

Hunter, M.L., 1996, *Fundamentals of Conservation Biology*, Oxford: Blackwell Science.

Inyang, E., 2005, *Towards Effective and Efficient Rumpi Project* [unpublished].

Kakonge, J.O., 1993, 'Constraints on Implementing Environmental Impact Assessments in Africa', *Environmental Impact Assessment Review*, Vol. 13, pp. 299–308.

Kakonge, J.O., 1994, 'Monitoring of Environmental Impact Assessments in Africa', *Environmental Impact Assessment Review*, Vol. 14, pp. 295–304.

Kakonge, J.O., 1998, 'EIA and Good Governance: Issues and Lessons from Africa', *Environmental Impact Assessment Review*, Vol.18, pp. 289–305.

Martin, P., 1990b, *First Steps to Sustainability: The School Curriculum and the Environment*, Surrey: WWF.

Patterson, T.E., 1993, *The American Democracy*, New York: McGraw-Hill.

SAIEA, 2004, *Environmental Impact Assessment in Southern Africa: Summary Report*, Windhoek: Southern African Institute for Environmental Assessment.

Schneider, H., Vinke, J. and Weekes-Vagliani, W., eds., 1993, *Environmental Education: An Approach to Sustainable Development*, Paris: Organisation for Economic Co-operation and Development, pp. 147–57.

Shimada, K., 2004, 'The Legacy of the Kyoto Protocol: Its Role as the Rulebook for an International Climate Framework', *International Review for Environmental Strategies*, Vol. 5, No, 1, pp. 3–14, Kamiyamaguchi: Institute for Global Environmental Strategies.

Struhsaker, T.T., 1998, *Ecology of an African rain forest: Logging in Kibale and the Conflict between Conservation and Exploitation*, Gainesville: University Press of Florida.

Tarr, P., 'EIA in Southern Africa: Summary and Future Focus', in P. Tarr, ed., *Environmental Impact Assessment in Southern Africa*, Windhoek: Southern African Institute for Environmental Assessment, pp. 329–337, 2003.

Umozurike, U. O., 1995, *Introduction to International Law*, Ibadan: Spectrum.

United Nations, 2005, *Kyoto Protocol to the United Nations Framework Convention on Climate Change*, United Nations University Press.

Weaver, A.V.B., Greling, T. and Wilgen, Van Wilgen, B.W., 1996, 'Logistics and Team Management of a Large Environmental Impact Assessment: Proposed Dune Mining at St. Lucia, South Africa', *Environmental Impact Assessment Review*, Vol. 16, pp. 103–113.

World Bank, 1992, *The World Development Report 1992*, Washington DC: Oxford.

WWF, 1988, *A Common Purpose: Environmental Education in the School Curriculum*, Surrey: WWF.

www.ingramcontent.com/pod-product-compliance
Lightning Source LLC
Chambersburg PA
CBHW072119020426
42334CB00018B/1653